United States Nuclear Regulatory Commission

Protecting People and the Environment

NUREG-0847
Supplement 26

Safety Evaluation Report

Related to the Operation of Watts Bar Nuclear Plant, Unit 2

Docket Number 50-391

Tennessee Valley Authority

Office of Nuclear Reactor Regulation

AVAILABILITY OF REFERENCE MATERIALS
IN NRC PUBLICATIONS

NRC Reference Material

As of November 1999, you may electronically access NUREG-series publications and other NRC records at NRC's Public Electronic Reading Room at http://www.nrc.gov/reading-rm.html. Publicly released records include, to name a few, NUREG-series publications; *Federal Register* notices; applicant, licensee, and vendor documents and correspondence; NRC correspondence and internal memoranda; bulletins and information notices; inspection and investigative reports; licensee event reports; and Commission papers and their attachments.

NRC publications in the NUREG series, NRC regulations, and Title 10, "Energy," in the *Code of Federal Regulations* may also be purchased from one of these two sources.

1. The Superintendent of Documents
   U.S. Government Printing Office
   Mail Stop SSOP
   Washington, DC 20402–0001
   Internet: bookstore.gpo.gov
   Telephone: 202-512-1800
   Fax: 202-512-2250
2. The National Technical Information Service
   Springfield, VA 22161–0002
   www.ntis.gov
   1–800–553–6847 or, locally, 703–605–6000

A single copy of each NRC draft report for comment is available free, to the extent of supply, upon written request as follows:
Address: U.S. Nuclear Regulatory Commission
         Office of Administration
         Publications Branch
         Washington, DC 20555-0001
E-mail: DISTRIBUTION.RESOURCE@NRC.GOV
Facsimile: 301–415–2289

Some publications in the NUREG series that are posted at NRC's Web site address http://www.nrc.gov/reading-rm/doc-collections/nuregs are updated periodically and may differ from the last printed version. Although references to material found on a Web site bear the date the material was accessed, the material available on the date cited may subsequently be removed from the site.

Non-NRC Reference Material

Documents available from public and special technical libraries include all open literature items, such as books, journal articles, transactions, *Federal Register* notices, Federal and State legislation, and congressional reports. Such documents as theses, dissertations, foreign reports and translations, and non-NRC conference proceedings may be purchased from their sponsoring organization.

Copies of industry codes and standards used in a substantive manner in the NRC regulatory process are maintained at—
        The NRC Technical Library
        Two White Flint North
        11545 Rockville Pike
        Rockville, MD 20852–2738

These standards are available in the library for reference use by the public. Codes and standards are usually copyrighted and may be purchased from the originating organization or, if they are American National Standards, from—
        American National Standards Institute
        11 West 42nd Street
        New York, NY 10036–8002
        www.ansi.org
        212–642–4900

United States Nuclear Regulatory Commission

Protecting People and the Environment

NUREG-0847
Supplement 26

Safety Evaluation Report

Related to the Operation of Watts Bar Nuclear Plant, Unit 2

Docket Number 50-391

Tennessee Valley Authority

Manuscript Completed: May 2013
Date Published: June 2013

Office of Nuclear Reactor Regulation

ABSTRACT

This report supplements the safety evaluation report (SER), NUREG-0847 (June 1982), Supplement No. 25 (November 2011, Agencywide Documents Access and Management System (ADAMS) Accession No. ML12011A024), with respect to the application filed by the Tennessee Valley Authority (TVA), as applicant and owner, for a license to operate Watts Bar Nuclear Plant (WBN) Unit 2 (Docket No 50-391).

In its SER and Supplemental SER (SSER) Nos. 1 through 20 issued by the Office of Nuclear Reactor Regulation (NRR) of the U.S. Nuclear Regulatory Commission (NRC), the NRC staff documented its safety evaluation and determination that WBN Unit 1 met all applicable regulations and regulatory guidance. Based on satisfactory findings from all applicable inspections, on February 7, 1996, the NRC issued a full-power operating license (OL) to WBN Unit 1, authorizing operation up to 100-percent power.

In SSER 21, the NRC staff addressed TVA's application for a license to operate WBN Unit 2, and provided information regarding the status of the items remaining to be resolved, which were outstanding at the time that TVA deferred construction of WBN Unit 2, and were not evaluated and resolved as part of the licensing of WBN Unit 1. Beginning with SSER 22, the NRC staff documented its ongoing evaluation and closure of open items in support of TVA's application for a license to operate WBN Unit 2.

In this and future SSERs, the NRC staff continues its documentation of its review of open items in support of TVA's application for an operating license for WBN Unit 2.

TABLE OF CONTENTS

ABBREVIATIONS

ABB	ASEA Brown Boveri
ABGTS	auxiliary building gas treatment system
AC or ac	alternating current
ACR	auxiliary control room
ADAMS	Agencywide Documents Access and Management System
AFW	auxiliary feedwater
ALARA	as low as reasonably achievable
ANSI	American National Standards Institute
ANS	American Nuclear Society
AOO	abnormal operational occurrence
AOP	abnormal operating procedure
APCSB	Auxiliary Power Conversion Systems Branch (of NRR)
ASME	American Society of Mechanical Engineers
AST	alternative source term
ASTM	American Society for Testing and Materials
AV	analysis volume
BEACON	Westinghouse Best Estimate Analyzer for Core Operations—Nuclear
BL	bulletin
BTP	Branch Technical Position
BWR	boiling-water reactor
CAP	corrective action program
CCP	centrifugal charging pump
CCS	component cooling system
CET	core exit thermocouple
cfm	cubic feet per minute
CFR	Code of Federal Regulations
CI	confirmatory issue
Ci	curie
CLB	current licensing basis
CMEB	Chemical Engineering Branch (of NRR)
CO_2	carbon dioxide
COMS	cold overpressure mitigation system
COT	channel operability test
CPU	central processor unit
CR	control room
CRC	Cyclic Redundancy Check
CSST	common station service transformer
CST	condensate storage tank
CT	current transformer
C_v	Charpy V-notch
CVCS	chemical and volume control system
DBA	design basis accident
DC or dc	direct current
DCF	dose conversion factor
DCN	design change notice
DCS	distributed control system

DEI	dose equivalent iodine-131
DF	decontamination factor
D/Q	deposition factor
DNB	departure from nucleate boiling
DNBR	departure from nucleate boiling ratio
EAB	exclusion area boundary
ECCS	emergency core cooling system
EDCR	Engineering Document Construction Release
EDG	emergency diesel generator
EGTS	emergency gas treatment system
EMC	electromagnetic compatibility
EMI/RFI	electromagnetic/radiofrequency interference
EOF	emergency operations facility
EOI	emergency operating instruction
EOP	emergency operating procedure
EPA	Environmental Protection Agency or electrical penetration assemblies
EPRI	Electric Power Research Institute
EQ	environmental qualification
ERCW	essential raw cooling water
ERD	Electronic Resources Division (Westinghouse)
ERDS	emergency response data system
ERFBS	electrical raceway fire barrier system
ESF	engineered safety feature
FHA	fuel handling accident
FHA	fire hazard analysis
FM	Factory Mutual
FPR	fire protection report
FSAR	final safety analysis report
FSSD	fire safe shutdown
FW	feedwater
GA	General Atomics
GDC	general design criterion/criteria
GL	generic letter
gpm	gallons per minute
GSI	generic safety issue
HEPA	high efficiency particulate air
HPFP	high pressure fire protection
HRCAR	high range containment air radiation
HTH	high-termpature heated treated
HVAC	heating, ventilation, and air conditioning
ICC	inadequate core cooling
ICRP	International Commission on Radiological Protection
ICS	integrated computer system
IE	Office of Inspection and Enforcement
IEB	Office of Inspection and Enforcement Bulletin
IEEE	Institute of Electrical and Electronics Engineers
IIS	in-core instrumentation system
IITA	in-core instrumentation thimble assembly
IN	Information Notice

IPE	individual plant examination
IPEEE	individual plant examination of external events
IPS	intake pumping station
JFD	joint frequency distribution
kHz	kilohertz
kV	kilovolt
kVA	kilovolt ampere
kW	kilowatt
LOCA	loss-of-coolant accident
LOOP	loss of offsite power
LPMS	loose part monitoring system
LPZ	low-population zone
LTOP	low-temperature overpressure protection
LWR	light-water reactor
MCC	motor control center
MCR	main control room
MHIF	multiple high impedence faults
MI	Mineral Insulated
MIC	microbiologically induced corrosion
MSIV	main steam isolation valve
MSLB	main steam line break
MSO	multiple spurious operation
MTEB	Materials Engineering Branch (of NRR)
MTP	maintenance and test panel
MVA	megavolt-ampere
MWt	megawatts thermal
NEI	Nuclear Energy Institute
NFPA	National Fire Protection Association
NGDC	New Generation Development and Construction
NPP	Nuclear Performance Plan
NP-REP	Nuclear Power Radiological Emergency Plan
NRC	Nuclear Regulatory Commission
NRR	Office of Nuclear Reactor Regulation
NSSS	nuclear steam supply system
NUREG	report prepared by NRC staff
OBE	operating basis earthquake
ODCM	Offsite Dose Calculation Manual
OL	operating license
OMA	operator manual action
OSG	original steam generator
PAD	performance analysis and design
PAMS	postaccident monitoring system
PDR	public document room
PMF	probable maximum flood
PORV	power-operated relief valve
ppm	parts per million
PRT	pressurizer relief tank
PTLR	Pressure and Temperature Limits Report
PWR	pressurized-water reactor

QA	quality assurance
RAI	request for additional information
RBPVS	reactor building purge ventilation system
RCCA	rod cluster control assembly
RCB	reactor coolant pressure boundary
RCP	reactor cooling pump
RCS	reactor coolant system
RCW	raw cooling water
RES	radiant energy shield
RG	Regulatory Guide
RHR	residual heat removal
RPM	radiation protection manager
RPV	reactor pressure vessel
RV	reactor vessel
RVI	reactor vessel internals
RWST	refueling water storage tank
scfm	standard cfm
SDD	software design description
SDOE	Secure Development and Operational Environment
SE	safety evaluation
SER	safety evaluation report, NUREG-0847, dated June 1982
SFP	spent fuel pool
SG	steam generator
SIS	safety injection system
SP	special program
SPM	software program manual
SPND	self-powered neutron detector
SPS	signal processing system
SRO	senior reactor operator
SRP	Standard Review Plan, NUREG-0800
SRS	software requirements specification
SSC	systems, structures, and components
SSE	safe shutdown earthquake
SSER	Supplemental SER
Std.	Standard
SV	safety valve
SysRS	System Requirements Specification
THR	total heat release rate
TI	Technical Instruction
TID	total integrated dose
TMI	Three Mile Island
TPBAR	tritium production burnable absorber rod
TS	technical specification
TSTF	Technical Specification Task Force
TVA	Tennessee Valley Authority
UFSAR	Updated FSAR
UHS	ultimate heat sink
UL	Underwriters Laboratories
V	volt

VAC	volt alternating current
VDC	volt direct current
V&V	verification and validation
VCT	volume control tank
VPA	ventilation and purge air
WBA	Web-based ADAMS
WBN	Watts Bar Nuclear Plant
WCAP	Westinghouse Commercial Atomic Power (report)
WEC	Westinghouse Electric Corporation
WINCISE	Westinghouse INCore Information, Surveillance, and Engineering system
χ/Q	atmospheric dispersion estimate

1 INTRODUCTION AND DISCUSSION

1.1 Introduction

The Watts Bar Nuclear Plant (WBN or Watts Bar) is owned by the Tennessee Valley Authority (TVA) and is located in southeastern Tennessee approximately 50 miles northeast of Chattanooga. The facility consists of two Westinghouse-designed four-loop pressurized-water reactors (PWRs) within ice condenser containments.

In June 1982, the Nuclear Regulatory Commission staff (NRC staff) issued safety evaluation report (SER), NUREG-0847, "Safety Evaluation Report related to the operation of Watts Bar Nuclear Plant Units 1 and 2," regarding TVA's application for licenses to operate WBN Units 1 and 2. In SER Supplements (SSERs) 1 through 20, the NRC staff concluded that WBN Unit 1 met all applicable regulations and regulatory guidance and on February 7, 1996, the NRC issued an operating license (OL) to Unit 1. TVA did not complete WBN Unit 2, and the NRC did not make conclusions regarding it.

On March 4, 2009, TVA submitted an updated application in support of its request for an OL for WBN Unit 2, pursuant to Title 10 of the *Code of Federal Regulations* (10 CFR), Part 50, "Domestic Licensing of Production and Utilization Facilities."

In SSER 21, the NRC staff provided information regarding the status of the WBN Unit 2 items that remain to be resolved, which were outstanding at the time that TVA deferred construction of Unit 2, and which were not evaluated and resolved as part of the licensing of WBN Unit 1. In SSER 22, the NRC staff began the documentation of its evaluation and closure of open items in support of TVA's application for a license to operate WBN Unit 2.

In this and future SSERs, the NRC staff will continue the documentation of its evaluation and closure of open items in support of TVA's application.

The format of this document is consistent with the format and scope outlined in the "Standard Review Plan for the Review of Safety Analysis Reports for Nuclear Power Plants: LWR [Light-Water Reactor] Edition (NUREG-0800)," dated July 1981 (SRP, NUREG-0800). The NRC staff added additional chapters to address the overall assessment of the facility, Nuclear Performance Plan issues, and other generic regulatory topics.

Each of the sections and appendices of this supplement is numbered the same as the SER section that is being updated, and the discussions are supplementary to, and not in lieu of, the discussion in the SER, unless otherwise noted. For example, Appendix E continues to list the principal contributors to the SSER. However, the chronology of the safety review correspondence previously provided in Appendix A has been discontinued, and a reference is provided instead to the NRC's Agencywide Documents Access and Management System (ADAMS) or the Public Document Room (PDR). Public correspondence exchanged between the NRC and TVA is available through ADAMS or the PDR. References listed as "not publicly available" in the SSER contain proprietary information and have been withheld from public disclosure in accordance with 10 CFR 2.390.

Appendix HH includes an Action Items Table. This table provides a status of all the open items, confirmatory issues, and proposed license conditions that must be resolved prior to completion of an NRC finding of reasonable assurance on the OL application for WBN Unit 2. The NRC staff will maintain the Action Items Table and revise Appendix HH in future SSERs, and add new appendices, as necessary.

The NRC's ADAMS is the agency's official recordkeeping system. ADAMS has the full text of regulatory and technical documents and reports written by NRC, NRC contractors, or NRC licensees. Documents include NRC regulatory guides, NUREG-series reports, correspondence, inspection reports, and others. These documents are assigned accession numbers and are searchable and accessible in ADAMS. Documents are released periodically during the day in the ADAMS PUBLIC/Legacy Interface Combined (ADAMS PUBLIC) and Web-based ADAMS (WBA) interfaces; they are released once a day in Web-based Publicly Available Records System (PARS). These documents in full text can be searched using ADAMS accession numbers or specific fields and parameters such as docket number and document dates.

More information regarding ADAMS and help for accessing documents may be obtained on the NRC Public Web site at http://www.nrc.gov/reading-rm/adams/faq.html#1.

All WBN documents may be accessed using WBN docket numbers 05000390 and 05000391 for Units 1 and 2, respectively.

The WBN Unit 2 Project Manager is Justin C. Poole, who may be contacted by calling (301) 415-2048, by e-mail to Justin.Poole@nrc.gov, or by writing to the following address:

> Mr. Justin Poole
> U.S. Nuclear Regulatory Commission
> Mail Stop O-8G9A
> Washington, D.C. 20555

1.7 Summary of Outstanding Issues

The NRC staff documented its previous review and conclusions regarding the OL application for WBN Unit 1 in the SER (NUREG-0847, dated June 1982) and its supplements 1 through 20. Based on these reviews, the NRC staff issued an OL for WBN Unit 1 in 1996. In the SER and SSERs 1 through 20, the NRC staff also reviewed and approved certain topics for WBN Unit 2, though no final conclusions were made regarding an OL for WBN Unit 2. To establish the remaining scope and the regulatory framework for the NRC staff's review of an OL for WBN Unit 2, the NRC staff reviewed the SER and SSERs 1 thorough 20. Based on this review, the NRC staff identified "resolved" topics (i.e., out of scope for review) and "open" topics (i.e., in scope for NRC staff review) for WBN Unit 2. Where it was not clear whether the SER topic applied to Unit 2 or not, the NRC staff conservatively identified it as "open" pending further evaluation. It should be noted that these were not technical evaluations of each topic; rather, it was a status review to determine whether the topic was "open" or "resolved." The NRC staff documented this evaluation in SSER 21 as the baseline for resumption of the review of the OL application for Unit 2. Thus, SSER 21 reflects the status of the NRC staff's review of WBN Unit 2 up to 1995. The NRC staff notes that a subsequent, more detailed assessment may find some topics conservatively identified in the initial assessment as "open" that should be redefined as "closed." Conversely, the NRC staff notes that there may be circumstances that could result in the need to reopen some previously closed topic areas that may have been adequately documented and that are considered closed in SSER 21. Such cases will be identified by a footnote in future SSERs to document that previous "open" topics have been recategorized as "closed" without requiring further review, or vice versa.

The SER and SSERs 1 through 20 evaluated the changes to the FSAR until Amendment 91. FSAR Amendment 91 was the initial licensing basis for WBN Unit 1. At this time, the FSAR was applicable to both Units 1 and 2. As part of its updated OL application for WBN Unit 2, TVA split the FSAR Amendment 91 into two separate FSARs for WBN Units 1 and 2. TVA has submitted WBN Unit 2 FSAR Amendments 92 through 109 to address the "open" topics in support of its OL application for WBN Unit 2. These FSAR amendments reflect changes that have occurred since 1995. These FSAR amendments are currently under NRC staff review. The NRC staff's review of these FSAR changes is documented in SSER 22 and subsequent supplements.

Additional general topics (e.g., financial qualifications that were not included in SSER 21, but that should be resolved prior to issuance of an OL) are also identified in SSER 22 and subsequent supplements.

SSER 21 initially provided the table below documenting the status of each SER topic. The relevant document in which the topic was last addressed is shown in parenthesis. This table will be maintained in this and future supplements to reflect the updated status of review for each topic.

ISSUE STATUS TABLE

	Issue	Status		Section	Note
(1)	Site Envelope			2	
(2)	Geography and Demography	Resolved	(SSER 22)	2.1	
(3)	Site Location and Description	Resolved	(SER) (SSER 22)	2.1.1	3
(4)	Exclusion Area Authority and Control	Resolved	(SER) (SSER 22)	2.1.2	3
(5)	Population Distribution	Resolved	(SER) (SSER 22)	2.1.3	
(6)	Conclusions	Resolved	(SER) (SSER 22)	2.1.4	
(7)	Nearby Industrial, Transportation, and Military Facilities	Resolved	(SSER 22)	2.2	
(8)	Transportation Routes	Resolved	(SER) (SSER 22)	2.2.1	
(9)	Nearby Facilities	Resolved	(SER) (SSER 22)	2.2.2	
(10)	Conclusions	Resolved	(SER) (SSER 22)	2.2.3	
(11)	Meteorology	Resolved	(SER) (SSER 22)	2.3	
(12)	Regional Climatology	Resolved	(SER) (SSER 22)	2.3.1	
(13)	Local Meteorology	Resolved	(SER) (SSER 22)	2.3.2	
(14)	Onsite Meteorological Measurements Program	Resolved	(SER) (SSER 22) (SSER 25)	2.3.3	

	Issue	Status		Section	Note
(15)	Short-Term (Accident) Atmospheric Diffusion Estimates	Resolved	(SER) (SSER 14) (SSER 22)	2.3.4	
(16)	Long-Term (Routine) Diffusion Estimates	Resolved	(SER) (SSER 14) (SSER 22)	2.3.5	
(17)	Hydrologic Engineering			2.4	
(18)	Introduction	Resolved	(SER)	2.4.1	
(19)	Hydrologic Description	Resolved	(SER)	2.4.2	
(20)	Flood Potential	Resolved	(SER)	2.4.3	
(21)	Local Intense Precipitation in Plant Area	Resolved	(SER)	2.4.4	1
(22)	Roof Drainage	Resolved	(SER)	2.4.5	1
(23)	Ultimate Heat Sink	Resolved	(SER)	2.4.6	
(24)	Groundwater	Resolved	(SER)	2.4.7	1
(25)	Design Basis for Subsurface Hydrostatic Loading	Resolved	(SER) (SSER 3)	2.4.8	
(26)	Transport of Liquid Releases	Resolved	(SER) (SSER 22)	2.4.9	2
(27)	Flooding Protection Requirements	Open (NRR)	(SER) (SSER 24)	2.4.10	
(28)	Geological, Seismological, and Geotechnical Engineering	Resolved	(SER) (SSER 24)	2.5	
(29)	Geology	Resolved	(SER)	2.5.1	
(30)	Seismology	Resolved	(SER)	2.5.2	
(31)	Surface Faulting	Resolved	(SER)	2.5.3	
(32)	Stability of Subsurface Materials and Foundations	Resolved	(SER) (SSER 3) (SSER 9) (SSER 11)	2.5.4	
(33)	Stability of Slopes	Resolved	(SER)	2.5.5	
(34)	Embankments and Dams	Resolved	(SER) (SSER 22)	2.5.6	
(35)	References		(SER) (SSER 22)	2.6	
(36)	Design Criteria - Structures, Components, Equipment, and Systems			3	
(37)	Introduction			3.1	
(38)	Conformance With General Design Criteria	Resolved	(SER)	3.1.1	
(39)	Conformance With Industry Codes and Standards	Resolved	(SER)	3.1.2	
(40)	Classification of Structures, Systems and Components	Resolved	(SSER 14) (SSER 22)	3.2	

	Issue	Status		Section	Note
(41)	Seismic Classifications	Resolved	(SER) (SSER 3) (SSER 5) (SSER 6) (SSER 8)	3.2.1	
(42)	System Quality Group Classification	Open (NRR)	(SER) (SSER 3) (SSER 6) (SSER 7) (SSER 9) (SSER 22)	3.2.2	
(43)	Wind and Tornado Loadings			3.3	
(44)	Wind Loading	Resolved	(SER)	3.3.1	
(45)	Tornado Loading	Resolved	(SER)	3.3.2	
(46)	Flood Level (Flood) Design			3.4	
(47)	Flood Protection	Resolved	(SER)	3.4.1	
(48)	Missile Protection			3.5	
(49)	Missile Selection and Description	Resolved	(SER) (SSER 9) (SSER 14) (SSER 22)	3.5.1	
(50)	Structures, Systems, and Components to be Protected from Externally Generated Missiles	Resolved	(SER) (SSER 2) (SSER 22)	3.5.2	
(51)	Barrier Design Procedures	Resolved	(SER)	3.5.3	
(52)	Protection Against the Dynamic Effects Associated with the Postulated Rupture of Piping	Open (NRR)	(SER) (SSER 6) (SSER 11)	3.6	
(53)	Plant Design for Protection Against Postulated Piping Failures in Fluid System Outside Containment	Resolved	(SER) (SSER 14) (SSER 22)	3.6.1	
(54)	Determination of Break Locations and Dynamic Effects Associated with the Postulated Rupture of Piping	Resolved	(SER) (SSER 14) (SSER 22)	3.6.2	3
(55)	Leak-Before-Break Evaluation Procedures	Resolved	(SSER 5) (SSER 12) (SSER 22) (SSER 24)	3.6.3	
(56)	Seismic Design	Resolved	(SER) (SSER 6)	3.7	2
(57)	Seismic Input	Resolved	(SER) (SSER 6) (SSER 9) (SSER 16)	3.7.1	2

	Issue	Status		Section	Note
(58)	Seismic Analysis	Resolved	(SER) (SSER 6) (SSER 8) (SSER 11) (SSER 16)	3.7.2	2
(59)	Seismic Subsystem Analysis	Resolved	(SER) (SSER 6) (SSER 7) (SSER 8) (SSER 9) (SSER 12) (SSER 22)	3.7.3	
(60)	Seismic Instrumentation	Resolved	(SER)	3.7.4	1
(61)	Design of Seismic Category I Structures	Resolved	(SER) (SSER 9)	3.8	2
(62)	Steel Containment	Resolved	(SER) (SSER 3)	3.8.1	
(63)	Concrete and Structural Steel Internal Structures	Resolved	(SER) (SSER 7)	3.8.2	
(64)	Other Seismic Category I Structures	Open (NRR)	(SER) (SSER 14) (SSER 16)	3.8.3	
(65)	Foundations	Resolved	(SER)	3.8.4	
(66)	Mechanical Systems and Components	Resolved	(SER)	3.9	
(67)	Special Topics for Mechanical Components	Resolved	(SER) (SSER 6) (SSER 13) (SSER 22)	3.9.1	
(68)	Dynamic Testing and Analysis of Systems, Components, and Equipment	Resolved	(SER) (SSER 14) (SSER 22)	3.9.2	
(69)	ASME Code Class 1, 2, and 3 Components, Component Structures, and Core Support Structures	Resolved	(SER) (SSER 3) (SSER 4) (SSER 6) (SSER 7) (SSER 8) (SSER 15) (SSER 22)	3.9.3	
(70)	Control Rod Drive Systems	Resolved	(SER)	3.9.4	
(71)	Reactor Pressure Vessel Internals	Open	(SER) (SSER 23)	3.9.5	

	Issue	Status		Section	Note
(72)	Inservice Testing of Pumps and Valves	Open (NRR)	(SER) (SSER 5) (SSER 12) (SSER 14) (SSER 18) (SSER 20) (SSER 22)	3.9.6	
(73)	Seismic and Dynamic Qualification of Seismic Category I Mechanical and Electrical Equipment	Resolved	(SER) (SSER 1) (SSER 3) (SSER 4) (SSER 5) (SSER 6) (SSER 8) (SSER 9) (SSER 23)	3.10	
(74)	Environmental Qualification of Mechanical and Electrical Equipment	Open (NRR)	(SSER 15) (SSER 22)	3.11	
(75)	Threaded Fasteners — ASME Code Class 1, 2, and 3	Resolved	(SSER 22)	3.13	
(76)	Reactor			4	
(77)	Introduction		(SER) (SSER 23)	4.1	
(78)	Fuel System Design	Open (NRR)	(SSER 23)	4.2	
(79)	Description	Resolved	(SER) (SSER 13) (SSER 23)	4.2.1	
(80)	Thermal Performance	Open (NRR)	(SER) (SSER 2) (SSER 23)	4.2.2	
(81)	Mechanical Performance	Resolved	(SER) (SSER 2) (SSER 10) (SSER 13) (SSER 23)	4.2.3	
(82)	Surveillance	Resolved	(SER) (SSER 2) (SSER 23)	4.2.4	
(83)	Fuel Design Considerations	Resolved	(SER) (SSER 23)	4.2.5	
(84)	Nuclear Design	Resolved	(SSER 23)	4.3	
(85)	Design Basis	Resolved	(SER) (SSER 13) (SSER 23)	4.3.1	
(86)	Design Description	Resolved	(SER) (SSER 13) (SSER 15) (SSER 23)	4.3.2	

	Issue	Status		Section	Note
(87)	Analytical Methods	Resolved	(SER) (SSER 23)	4.3.3	
(88)	Summary of Evaluation Findings	Resolved	(SER) (SSER 23)	4.3.4	
(89)	Thermal-Hydraulic Design	Resolved	(SSER 23)	4.4	
(90)	Performance in Safety Criteria	Resolved	(SER) (SSER 23)	4.4.1	
(91)	Design Bases	Resolved	(SER) (SSER 12) (SSER 23)	4.4.2	
(92)	Thermal-Hydraulic Design Methodology	Resolved	(SER) (SSER 6) (SSER 8) (SSER 12) (SSER 13) (SSER 16) SE dated 6/13/89 (SSER 23)	4.4.3	
(93)	Operating Abnormalities	Resolved	(SER) (SSER 13) (SSER 23)	4.4.4	
(94)	Loose Parts Monitoring System	Resolved	(SER) (SSER 3) (SSER 5) (SSER 16) (SSER 23)	4.4.5	
(95)	Thermal-Hydraulic Comparison	Resolved	(SER) (SSER 23)	4.4.6	
(96)	N-1 Loop Operation	Resolved	(SER) (SSER 23)	4.4.7	
(97)	Instrumentation for Inadequate Core Cooling Detection (TMI Action Item II.F.2)	Open (NRR)	(SER) (SSER 10) (SSER 23)	4.4.8	
(98)	Summary and Conclusion	Open (NRR)	(SER) (SSER 23)	4.4.9	
(99)	Reactor Materials			4.5	
(100)	Control Rod Drive Structural Materials	Resolved	(SER)	4.5.1	1
(101)	Reactor Internals and Core Support Materials	Resolved	(SER)	4.5.2	
(102)	Functional Design of Reactivity Control Systems	Resolved	(SER) (SSER 23)	4.6	
(103)	Reactor Coolant System and Connected Systems			5	
(104)	Summary Description	Resolved	(SER) (SSER 5) (SSER 6)	5.1	2

	Issue	Status		Section	Note
(105)	Integrity of Reactor Coolant Pressure Boundary			5.2	
(106)	Compliance with Codes and Code Cases	Resolved	(SER) (SSER 22)	5.2.1	
(107)	Overpressurization Protection	Resolved	(SER) (SSER 2) (SSER 15) (SSER 24)	5.2.2	
(108)	Reactor Coolant Pressure Boundary Materials	Resolved	(SER) (SSER 22)	5.2.3	
(109)	Reactor Coolant System Pressure Boundary Inservice Inspection and Testing	Open (NRR)	(SER) (SSER 10) (SSER 12) (SSER 15) (SSER 16) (SSER 23)	5.2.4	
(110)	Reactor Coolant Pressure Boundary Leakage Detection	Resolved	(SER) (SSER 9) (SSER 11) (SSER 12) (SSER 22)	5.2.5	
(111)	Reactor Vessel			5.3	
(112)	Reactor Vessel Materials	Resolved	(SER) (SSER 11) (SSER 14) (SSER 22) (SSER 25)	5.3.1	
(113)	Pressure-Temperature Limits	Resolved	(SER) (SSER 16) (SSER 22) (SSER 25)	5.3.2	
(114)	Reactor Vessel Integrity	Open (NRR)	(SER) (SSER 22)	5.3.3	
(115)	Component and Subsystem Design			5.4	
(116)	Reactor Coolant Pumps	Resolved	(SER) (SSER 22)	5.4.1	2
(117)	Steam Generators	Resolved	(SER) (SSER 1) (SSER 4) (SSER 22)	5.4.2	
(118)	Residual Heat Removal System	Resolved	(SER) (SSER 2) (SSER 5) (SSER 10) (SSER 11) (SSER 23)	5.4.3	
(119)	Pressurizer Relief Tank	Resolved	(SER) (SSER 22)	5.4.4	

	Issue	Status		Section	Note
(120)	Reactor Coolant System Vents (TMI Action Item II.B.1)	Open (Inspection)	(SER) (SSER 2) (SSER 5) (SSER 12) (SSER 23)	5.4.5	
(121)	Engineered Safety Features			6	
(122)	Engineered Safety Feature Materials			6.1	
(123)	Metallic Materials	Open	(SER) (SSER 23)	6.1.1	
(124)	Organic Materials	Resolved	(SER) (SSER 22)	6.1.2	
(125)	Postaccident Emergency Cooling Water Chemistry	Resolved	(SER) (SSER 22)	6.1.3	
(126)	Containment Systems			6.2	
(127)	Containment Functional Design	Resolved	(SER) (SSER 3) (SSER 5) (SSER 7) (SSER 12) (SSER 14) (SSER 15) (SSER 22)	6.2.1	
(128)	Containment Heat Removal Systems	Resolved	(SER) (SSER 7) (SSER 22)	6.2.2	
(129)	Secondary Containment Functional Design	Resolved	(SER) (SSER 18) (SSER 22)	6.2.3	
(130)	Containment Isolation Systems	Resolved	(SER) (SSER 3) (SSER 5) (SSER 7) (SSER 12) (SSER 22)	6.2.4	
(131)	Combustible Gas Control Systems	Resolved	(SER) (SSER 4) (SSER 5) (SSER 8) (SSER 22)	6.2.5	
(132)	Containment Leakage Testing	Open (NRR)	(SER) (SSER 4) (SSER 5) (SSER 19) (SSER 22)	6.2.6	
(133)	Fracture Prevention of Containment Pressure Boundary	Resolved	(SER) (SSER 4) (SSER 23)	6.2.7	1
(134)	Emergency Core Cooling System	Resolved	(SER)	6.3	1

	Issue	Status		Section	Note
(135)	System Design	Open (NRR)	(SER)	6.3.1	
			(SSER 6)		
			(SSER 7)		
			(SSER 11)		
(136)	Evaluation	Resolved	(SER)	6.3.2	1
			(SSER 5)		
(137)	Testing	Open (NRR)	(SER)	6.3.3	
			(SSER 2)		
			(SSER 9)		
(138)	Performance Evaluation	Resolved	(SER)	6.3.4	
(139)	Conclusions	Open (NRR)	(SER)	6.3.5	
(140)	Control Room Habitability	Resolved	(SER)	6.4	
			(SSER 5)		
			(SSER 11)		
			(SSER 16)		
			(SSER 18		
			(SSER 22)		
(141)	Engineered Safety Feature (ESF) Filter Systems			6.5	
(142)	ESF Atmosphere Cleanup System	Resolved	(SER)	6.5.1	
			(SSER 5)		
			(SSER 22)		
(143)	Fission Product Cleanup System	Resolved	(SER)	6.5.2	1
(144)	Fission Product Control System	Open (NRR)	(SER)	6.5.3	
			(SSER 22)		
(145)	Ice Condenser as a Fission Product Cleanup System	Resolved	(SER)	6.5.4	1
(146)	Inservice Inspection of Class 2 and 3 Components	Open (NRR)	(SER)	6.6	
			(SSER 10)		
			(SSER 12)		
			(SSER 15)		
			(SSER 23)		
(147)	Instrumentation and Controls			7	
(148)	Introduction			7.1	
(149)	General	Resolved	(SER)	7.1.1	
			(SSER 13)		
			(SSER 16)		
			(SSER 23)		
(150)	Comparison with Other Plants	Resolved	(SER)	7.1.2	1
			(SSER 23)		
(151)	Design Criteria	Resolved	(SER)	7.1.3	
			(SSER 4)		
			(SSER 15)		
			(SSER 23)		
(152)	Reactor Trip System	Resolved	(SER)	7.2	
(153)	System Description	Open (NRR)	(SER)	7.2.1	
			(SSER 13)		
			(SSER 15)		
			(SSER 23)		

Issue		Status		Section	Note
(154)	Manual Trip Switches	Resolved	(SER) (SSER 23)	7.2.2	1
(155)	Testing of Reactor Trip Breaker Shunt Coils	Resolved	(SER) (SSER 23)	7.2.3	1
(156)	Anticipatory Trips	Resolved	(SER) (SSER 23)	7.2.4	
(157)	Steam Generator Water Level Trip	Resolved	(SER) (SSER 2) (SSER 14) (SSER 23)	7.2.5	
(158)	Conclusions	Resolved	(SER) (SSER 13) (SSER 23)	7.2.6	
(159)	Engineered Safety Features System	Open (NRR)	(SER) (SSER 13)	7.3	
(160)	System Description	Resolved	(SER) (SSER 13) (SSER 14) (SSER 23)	7.3.1	
(161)	Containment Sump Level Measurement	Resolved	(SER) (SSER 2) (SSER 23)	7.3.2	
(162)	Auxiliary Feedwater Initiation and Control	Resolved	(SER) (SSER 23)	7.3.3	1
(163)	Failure Modes and Effects Analysis	Resolved	(SER) (SSER 23)	7.3.4	
(164)	IE Bulletin 80-06	Resolved	(SER) (SSER 3) (SSER 23)	7.3.5	
(165)	Conclusions	Resolved	(SER) (SSER 13) (SSER 23)	7.3.6	
(166)	Systems Required for Safe Shutdown			7.4	
(167)	System Description	Resolved	(SER) (SSER 23)	7.4.1	
(168)	Safe Shutdown from Auxiliary Control Room	Resolved	(SER) (SSER 7) (SSER 23)	7.4.2	
(169)	Conclusions	Resolved	(SER) (SSER 23)	7.4.3	
(170)	Safety-Related Display Instrumentation			7.5	
(171)	Display Systems	Resolved	(SER) (SSER 23)	7.5.1	

	Issue	Status		Section	Note
(172)	Postaccident Monitoring System	Open (Inspection)	(SER) (SSER 7) (SSER 9) (SSER 14) (SSER 15) (SSER 23) (SSER 25)	7.5.2	
(173)	IE Bulletin 79-27	Open (Inspection)	(SER) (SSER 23)	7.5.3	
(174)	Conclusions	Open (Inspection)	(SER)	7.5.4	
(175)	All Other Systems Required for Safety			7.6	
(176)	Loose Part Monitoring System	Resolved	(SER) (SSER 23) (SSER 24)	7.6.1	
(177)	Residual Heat Removal System Bypass Valves	Resolved	(SER) (SSER 23)	7.6.2	
(178)	Upper Head Injection Manual Control	Resolved	(SER) (SSER 23)	7.6.3	
(179)	Protection Against Spurious Actuation of Motor-Operated Valves	Resolved	(SER) (SSER 23)	7.6.4	
(180)	Overpressure Protection during Low Temperature Operation	Resolved	(SER) (SSER 4) (SSER 23)	7.6.5	
(181)	Valve Power Lockout	Resolved	(SER) (SSER 23)	7.6.6	
(182)	Cold Leg Accumulator Valve Interlocks and Position Indication	Resolved	(SER) (SSER 23)	7.6.7	
(183)	Automatic Switchover From Injection to Recirculation Mode	Resolved	(SER) (SSER 23)	7.6.8	
(184)	Conclusions	Resolved	(SER) (SSER 4)	7.6.9	
(185)	Control Systems Not Required for Safety			7.7	
(186)	System Description	Open (NRR)	(SER) (SSER 23) (SSER 24) (SSER 25)	7.7.1	
(187)	Safety System Status Monitoring System	Resolved	(SER) (SSER 7) (SSER 13) (SSER 23)	7.7.2	
(188)	Volume Control Tank Level Control System	Resolved	(SER) (SSER 23)	7.7.3	
(189)	Pressurizer and Steam Generator Overfill	Resolved	(SER) (SSER 23)	7.7.4	
(190)	IE Information Notice 79-22	Resolved	(SER) (SSER 23)	7.7.5	

	Issue	Status		Section	Note
(191)	Multiple Control System Failures	Resolved	(SER) (SSER 23)	7.7.6	
(192)	Conclusions	Resolved	(SER)	7.7.7	
(193)	Anticipated Transient Without Scram Mitigation System Actuation Circuitry (AMSAC)	Resolved	(SSER 9) (SSER 14) (SSER 23)	7.7.8	
(194)	NUREG-0737 Items	Resolved	(SER) (SSER 23)	7.8	
(195)	Relief and Safety Valve Position Indication (TMI Action Item II.D.3)	Open (Inspection)	(SER) (SSER 5) (SSER 14) (SSER 23)	7.8.1	
(196)	Auxiliary Feedwater System Initiation and Flow Indication (TMI Action Item II.E.1.2)	Open (Inspection)	(SER) (SSER 23)	7.8.2	
(197)	Proportional Integral Derivative Control Modification (TMI Action Item II.K.3.9)	Open (Inspection)	(SER) (SSER 23)	7.8.3	
(198)	Proposed Anticipatory Trip Modification (TMI Action Item II.K.3.10)	Resolved	(SER) (SSER 4) (SSER 23)	7.8.4	
(199)	Confirm Existence of Anticipatory Reactor Trip Upon Turbine Trip (TMI Action Item II.K.3.12)	Resolved	(SER) (SSER 23)	7.8.5	
(200)	Data Communication Systems		(SSER 23)	7.9	
(201)	Electric Power Systems			8	
(202)	General	Open (NRR)	(SER) (SSER 22) (SSER 24)	8.1	
(203)	Offsite Power System	Open (NRR)	(SER) (SSER 22)	8.2	
(204)	Compliance with GDC 5	Open (NRR)	(SER) (SSER 13) (SSER 22)	8.2.1	
(205)	Compliance with GDC 17	Open (NRR)	(SER) (SSER 2) (SSER 3) (SSER 13) (SSER 14) (SSER 15 (SSER 22)	8.2.2	
(206)	Compliance with GDC 18	Resolved	(SER) (SSER 22)	8.2.3	
(207)	Evaluation Findings	Open (NRR)	(SER) (SSER 22)	8.2.4	
(208)	Onsite Power Systems	Resolved	(SER) (SSER 10) (SSER 19) (SSER 22)	8.3	

	Issue	Status		Section	Note
(209)	Onsite AC Power System Compliance with GDC 17	Open (NRR)	(SER) (SSER 2) (SSER 7) (SSER 9) (SSER 10) (SSER 13) (SSER 14) (SSER 18) (SSER 20) (SSER 22)	8.3.1	
(210)	Onsite DC System Compliance with GDC 17	Open (NRR)	(SER) (SSER 2) (SSER 3) (SSER 13) (SSER 14) (SSER 22)	8.3.2	
(211)	Common Electrical Features and Requirements	Resolved	(SER) (SSER 2) (SSER 3) (SSER 7) (SSER 13) (SSER 14) (SSER 15) (SSER 16) (SSER 22)	8.3.3	
(212)	Evaluation Findings	Open (NRR)	(SER) (SSER 2) (SSER 3) (SSER 7) (SSER 13) (SSER 14) (SSER 15) (SSER 16) (SSER 22)	8.3.4	
(213)	Station Blackout	Open (NRR)	(SSER 22)	8.4	
(214)	Auxiliary Systems	Resolved	(SER) (SSER 10)	9	
(215)	Fuel Storage Facility			9.1	
(216)	New-Fuel Storage	Resolved	(SER)	9.1.1	1
(217)	Spent-Fuel Storage	Resolved	(SER) (SSER 5) (SSER 15) (SSER 16) (SSER 22)	9.1.2	
(218)	Spent Fuel Pool Cooling and Cleanup System	Open (NRR)	(SER) (SSER 11) (SSER 15) (SSER 23)	9.1.3	

	Issue	Status		Section	Note
(219)	Fuel-Handling System	Resolved	(SER) (SSER 3) (SSER 13) (SSER 22) (SSER 24)	9.1.4	
(220)	Water Systems			9.2	
(221)	Essential Raw Cooling Water and Raw Cooling Water System	Open (NRR)	(SER) (SSER 9) (SSER 10) (SSER 18) (SSER 23)	9.2.1	
(222)	Component Cooling System (Reactor Auxiliaries Cooling Water System)	Open (NRR)	(SER) (SSER 5) (SSER 23)	9.2.2	
(223)	Demineralized Water Makeup System	Resolved	(SER) (SSER 22)	9.2.3	
(224)	Potable and Sanitary Water Systems	Resolved	(SER) (SSER 9) (SSER 22)	9.2.4	
(225)	Ultimate Heat Sink	Open (NRR)	(SER) (SSER 23)	9.2.5	
(226)	Condensate Storage Facilities	Resolved	(SER) (SSER 12) (SSER 22)	9.2.6	
(227)	Process Auxiliaries			9.3	
(228)	Compressed Air System	Resolved	(SER) (SSER 22)	9.3.1	1
(229)	Process Sampling System	Resolved	(SER) (SSER 3) (SSER 5) (SSER 14) (SSER 16) (SSER 24)	9.3.2	
(230)	Equipment and Floor Drainage System	Resolved	(SER) (SSER 22)	9.3.3	3
(231)	Chemical and Volume Control System	Resolved	(SER) (SSER 22)	9.3.4	3
(232)	Heat Tracing		(SSER 22)	9.3.8	
(233)	Heating, Ventilation, and Air Conditioning Systems			9.4	
(234)	Control Room Area Ventilation System	Resolved	(SER) (SSER 9) (SSER 22)	9.4.1	
(235)	Fuel-Handling Area Ventilation System	Resolved	(SER) (SSER 22)	9.4.2	
(236)	Auxiliary Building and Radwaste Area Ventilation System	Resolved	(SER) (SSER 22)	9.4.3	
(237)	Turbine Building Area Ventilation System	Resolved	(SER) (SSER 22)	9.4.4	

	Issue	Status		Section	Note
(238)	Engineered Safety Features Ventilation System	Resolved	(SER) (SSER 9) (SSER 10) (SSER 11) (SSER 14) (SSER 16) (SSER 19) (SSER 22)	9.4.5	
(239)	Reactor Building Purge Ventilation System		(SSER 22)	9.4.6	
(240)	Containment Air Cooling System		(SSER 22)	9.4.7	
(241)	Condensate Demineralizer Waste Evaporator Building Environmental Control System		(SSER 22)	9.4.8	
(242)	Other Auxiliary Systems			9.5	
(243)	Fire Protection	Resolved	(SER) (SSER 10) (SSER 18) (SSER 19) (SSER 26)	9.5.1	3
(244)	Communications System	Resolved	(SER) (SSER 5)	9.5.2	1
(245)	Lighting System	Resolved	(SER) (SSER 22)	9.5.3	
(246)	Emergency Diesel Engine Fuel Oil Storage and Transfer System	Resolved	(SER) (SSER 5) (SSER 9) (SSER 10) (SSER 11) (SSER 12) (SSER 22)	9.5.4	2
(247)	Emergency Diesel Engine Cooling Water System	Resolved	(SER) (SSER 5) (SSER 11)	9.5.5	1
(248)	Emergency Diesel Engine Starting Systems	Resolved	(SER) (SSER 5) (SSER 10) (SSER 22)	9.5.6	2
(249)	Emergency Diesel Engine Lubricating Oil System	Resolved	(SER) (SSER 3) (SSER 5) (SSER 10) (SSER 22)	9.5.7	2
(250)	Emergency Diesel Engine Combustion Air Intake and Exhaust System	Resolved	(SER) (SSER 5) (SSER 10) (SSER 22)	9.5.8	2
(251)	Steam and Power Conversion System			10	

	Issue	Status		Section	Note
(252)	Summary Description	Resolved	(SER)	10.1	
(253)	Turbine Generator	Open (NRR)	(SER) (SSER 5)	10.2	
(254)	Turbine Generator Design	Resolved	(SER) (SSER 12) (SSER 22)	10.2.1	
(255)	Turbine Disc Integrity	Resolved	(SER) (SSER 23)	10.2.2	
(256)	Main Steam Supply System	Resolved	(SER)	10.3	
(257)	Main Steam Supply System (Up to and Including the Main Steam Isolation Valves)	Resolved	(SER) (SSER 19) (SSER 22)	10.3.1	
(258)	Main Steam Supply System	Resolved	(SER) (SSER 22)	10.3.2	2
(259)	Steam and Feedwater System Materials	Resolved	(SER) (SSER 22)	10.3.3	
(260)	Secondary Water Chemistry	Resolved	(SER) (SSER 5) (SSER 22)	10.3.4	
(261)	Other Features			10.4	
(262)	Main Condenser	Resolved	(SER) (SSER 9) (SSER 22)	10.4.1	
(263)	Main Condenser Evacuation System	Resolved	(SER) (SSER 22)	10.4.2	
(264)	Turbine Gland Sealing System	Resolved	(SER) (SSER 22)	10.4.3	
(265)	Turbine Bypass System	Resolved	(SER) (SSER 5) (SSER 22)	10.4.4	
(266)	Condenser Circulating Water System	Resolved	(SER) (SSER 22)	10.4.5	
(267)	Condensate Cleanup System	Open (NRR)	(SER) (SSER 22)	10.4.6	
(268)	Condensate and Feedwater Systems	Resolved	(SER) (SSER 14) (SSER 22)	10.4.7	
(269)	Steam Generator Blowdown System	Resolved	(SER) (SSER 22) (SSER 24)	10.4.8	
(270)	Auxiliary Feedwater System	Resolved	(SER) (SSER 14) (SSER 23) (SSER 24)	10.4.9	
(271)	Heater Drains and Vents	Resolved	(SSER 22)	10.4.10	
(272)	Steam Generator Wet Layup System	Resolved	(SSER 22)	10.4.11	
(273)	Radioactive Waste Management			11	

	Issue	Status		Section	Note
(274)	Summary Description	Resolved	(SER) (SSER 16) (SSER 24)	11.1	2
(275)	Liquid Waste Management	Resolved	(SER) (SSER 4) (SSER 16) (SSER 24)	11.2	
(276)	Gaseous Waste Management	Resolved	(SER) (SSER 8) (SSER 16) (SSER 24) (SSER 25)	11.3	
(277)	Solid Waste Management System	Resolved	(SER) (SSER 16) (SSER 24)	11.4	
(278)	Process and Effluent Radiological Monitoring and Sampling Systems	Resolved	(SER) (SSER 16) (SSER 20) (SSER 24)	11.5	
(279)	Evaluation Findings	Resolved	(SER) (SSER 8) (SSER 16)	11.6	
(280)	NUREG-0737 Items	Open (NRR)	(SER)	11.7	
(281)	Wide-Range Noble Gas, Iodine, and Particulate Effluent Monitors (TMI Action Items II.F.1(1) and II.F.1(2))	Open (Inspection)	(SER) (SSER 5) (SSER 6)	11.7.1	
(282)	Primary Coolant Outside Containment (TMI Action item III.D.1.1)	Open (NRR)	(SER) (SSER 5) (SSER 6) (SSER 10) (SSER 16)	11.7.2	
(283)	Radiation Protection			12	
(284)	General	Resolved	(SER) (SSER 10) (SSER 14) (SSER 24)	12.1	
(285)	Ensuring that Occupational Radiation Doses Are As Low As Reasonably Achievable (ALARA)	Resolved	(SER) (SSER 14) (SSER 24)	12.2	2
(286)	Radiation Sources	Resolved	(SER) (SSER 14) (SSER 24)	12.3	
(287)	Radiation Protection Design Features	Open (NRR)	(SER) (SSER 10) (SSER 14) (SSER 18) (SSER 24)	12.4	

	Issue	Status		Section	Note
(288)	Dose Assessment	Open (NRR)	(SER) (SSER 14) (SSER 24)	12.5	
(289)	Health Physics Program	Open (NRR)	(SER) (SSER 10) (SSER 14) (SSER 24)	12.6	
(290)	NUREG-0737 Items			12.7	
(291)	Plant Shielding (TMI Action Item II.B.2)	Open (NRR)	(SER) (SSER 14) (SSER 16) (SSER 24)	12.7.1	
(292)	High Range In-Containment Monitor (TMI Action Item II.F.1.(3))	Open (NRR)	(SER) (SSER 5)	12.7.2	
(293)	In-Plant Radioiodine Monitor (TMI Action Item II.D.3.3)	Open (NRR)	(SER) (SSER 16)	12.7.3	
(294)	Conduct of Operations			13	
(295)	Organization Structure of the Applicant	Resolved	(SER) (SSER 16) (SSER 22)	13.1	
(296)	Management and Technical Organization	Resolved	(SER)	13.1.1	
(297)	Corporate Organization and Technical Support	Resolved	(SER)	13.1.2	
(298)	Plant Staff Organization	Resolved	(SER) (SSER 8) (SSER 22) (SSER 25)	13.1.3	
(299)	Training			13.2	
(300)	Licensed Operator Training Program	Resolved	(SER) (SSER 9) (SSER 10) (SSER 22)	13.2.1	
(301)	Training for Non-licensed Personnel	Resolved	(SER)	13.2.2	
(302)	Emergency Preparedness Evaluation			13.3	
(303)	Introduction	Open (NRR)	(SER) (SSER 13) (SSER 20)	13.3.1	
(304)	Evaluation of the Emergency Plan	Open (NRR)	(SER) (SSER 13) (SSER 20) (SSER 22)	13.3.2	
(305)	Conclusions	Open (NRR)	(SER) (SSER 13) (SSER 20) (SSER 22)	13.3.3	

	Issue	Status		Section	Note
(306)	Review and Audit	Resolved	(SER) (SSER 8) (SSER 22)	13.4	
(307)	Plant Procedures	Resolved	(SER) (SSER 22)	13.5	
(308)	Administrative Procedures	Resolved	(SER) (SSER 22)	13.5.1	
(309)	Operating and Maintenance Procedures	Resolved	(SER) (SSER 9) (SSER 10) (SSER 22)	13.5.2	
(310)	NUREG-0737 Items	Resolved	(SER) (SSER 3) (SSER 16) (SSER 22)	13.5.3	
(311)	Physical Security Plan	Resolved	(SER) (SSER 1) (SSER 10) (SSER 15) (SSER 20) (SSER 22)	13.6	
(312)	Introduction	Resolved	(SSER 22)	13.6.1	
(313)	Summary of Application	Resolved	(SSER 22)	13.6.2	
(314)	Regulatory Basis	Resolved	(SSER 22)	13.6.3	
(315)	Technical Evaluation	Resolved	(SSER 22)	13.6.4	
(316)	Conclusions	Resolved	(SSER 22)	13.6.5	
(317)	Cyber Security Plan	Resolved	(SSER 24)	13.6.6	
(318)	Initial Test Program	Resolved	(SER) (SSER 3) (SSER 5) (SSER 7) (SSER 9) (SSER 10) (SSER 12) (SSER 14) (SSER 16) (SSER 18) (SSER 19) (SSER 23)	14	
(319)	Accident Analyses			15	
(320)	General Discussion	Resolved	(SER)	15.1	
(321)	Normal Operation and Anticipated Transients	Open (NRR)	(SER)	15.2	
(322)	Loss-of-Cooling Transients	Resolved	(SER) (SSER 13) (SSER 14) (SSER 24)	15.2.1	
(323)	Increased Cooling Inventory Transients	Resolved	(SER) (SSER 24)	15.2.2	

	Issue	Status		Section	Note
(324)	Change in Inventory Transients	Resolved	(SER) (SSER 18) (SSER 24)	15.2.3	
(325)	Reactivity and Power Distribution Anomalies	Open (NRR)	(SER) (SSER 4) (SSER 7) (SSER 13) (SSER 14) (SSER 24)	15.2.4	
(326)	Conclusions	Resolved	(SER) (SSER 4)	15.2.5	
(327)	Limiting Accidents	Resolved	(SER)	15.3	
(328)	Loss-of-Coolant Accident (LOCA)	Resolved	(SER) (SSER 12) (SSER 15) (SSER 24)	15.3.1	
(329)	Steamline Break	Resolved	(SER) (SSER 3) (SSER 14) (SSER 24)	15.3.2	
(330)	Feedwater System Pipe Break	Resolved	(SER) (SSER 14) (SSER 24)	15.3.3	
(331)	Reactor Coolant Pump Rotor Seizure	Resolved	(SER) (SSER 14) (SSER 24)	15.3.4	
(332)	Reactor Coolant Pump Shaft Break	Resolved	(SER) (SSER 14) (SSER 24)	15.3.5	
(333)	Anticipated Transients Without Scram	Resolved	(SER) (SSER 3) (SSER 5) (SSER 6) (SSER 10) (SSER 11) (SSER 12) (SSER 24)	15.3.6	
(334)	Conclusions	Resolved	(SER)	15.3.7	
(335)	Radiological Consequences of Accidents	Resolved	(SER) (SSER 15) (SSER 25)	15.4	
(336)	Loss-of-Coolant Accident	Resolved	(SER) (SSER 5) (SSER 9) (SSER 18) (SSER 25)	15.4.1	
(337)	Main Steamline Break Outside of Containment	Open (NRR)	(SER) (SSER 15) (SSER 25)	15.4.2	

	Issue	Status		Section	Note
(338)	Steam Generator Tube Rupture	Resolved	(SER) (SSER 2) (SSER 5) (SSER 12) (SSER 14) (SSER 15) (SSER 25)	15.4.3	
(339)	Control Rod Ejection Accident	Resolved	(SER) (SSER 15) (SSER 25)	15.4.4	
(340)	Fuel-Handling Accident	Resolved	(SER) (SSER 4) (SSER 15) (SSER 25)	15.4.5	
(341)	Failure of Small Line Carrying Coolant Outside Containment	Resolved	(SER) (SSER 25)	15.4.6	
(342)	Postulated Radioactive Releases as a Result of Liquid Tank Failures	Resolved	(SER) (SSER 25)	15.4.7	
(342a)	Postulated Waste Gas Decay Tank Rupture	Resolved	(SSER 25)	15.4.8	
(343)	NUREG-0737 Items			15.5	
(344)	Thermal Mechanical Report (TMI Action Item II.K.2.13)	Resolved	(SER) (SSER 4) (SSER 24)	15.5.1	
(345)	Voiding in the Reactor Coolant System during Transients (TMI Action Item II.K.2.17)	Resolved	(SER) (SSER 4) (SSER 24)	15.5.2	
(346)	Installation and Testing of Automatic Power-Operated Relief Valve Isolation System (TMI Action Item II.K.3.1) Report on Overall Safety Effect of Power-Operated Relief Valve Isolation System (TMI Action Item II.K.3.2)	Resolved	(SER) (SSER 5)	15.5.3	
(347)	Automatic Trip of Reactor Coolant Pumps (TMI Action Item II.K.3.5)	Resolved	(SER) (SSER 4) (SSER 16) (SSER 24)	15.5.4	
(348)	Small-Break LOCA Methods (II.K.3.30) and Plant-Specific Calculations (II.K.3.31)	Open (Inspection)	(SER) (SSER 4) (SSER 5) (SSER 16)	15.5.5	
(349)	Relative Risk of Low-Power Operation	Resolved	(SER)	15.6	
(350)	Technical Specification	Open (NRR)		16	
(351)	Quality Assurance			17	
(352)	General	Resolved	(SER)	17.1	
(353)	Organization	Resolved	(SER)	17.2	

	Issue	Status		Section	Note
(354)	Quality Assurance Program	Resolved	(SER) (SSER 2) (SSER 5) (SSER 10) (SSER 13) (SSER 15) (SSER 22)	17.3	
(355)	Conclusions	Resolved	(SER)	17.4	
(356)	Maintenance Rule			17.6	
(357)	Control Room Design Review			18	
(358)	General	Resolved	(SER) (SSER 5) (SSER 6) (SSER 15) (SSER 16) (SSER 22)	18.1	
(359)	Conclusions	Resolved	(SER) (SSER 16) (SSER 22)	18.2	
(360)	Report of the Advisory Committee on Reactor Safeguards	Open (NRR)	(SER)	19	
(361)	Common Defense and Security	Resolved	(SER)	20	
(362)	Financial Qualifications	Resolved	(SER)	21	
(363)	TVA Financial Qualifications for WBN Unit 2	Resolved	(SSER 22) (SSER 23)	21.1	
(364)	Foreign Ownership, Control, or Domination	Resolved	(SSER 22)	21.2	
(365)	Financial Protection and Indemnity Requirements			22	
(366)	General	Resolved	(SER)	22.1	
(367)	Preoperational Storage of Nuclear Fuel	Resolved	(SER)	22.2	
(368)	Operating Licenses	Open (NRR)	(SSER 22)	22.3	
(369)	Quality of Construction, Operational Readiness, and Quality Assurance Effectiveness			25	
(370)	Program for Maintenance and Preservation of the Licensing Basis for Units 1 and 2	Open (NRR)	(SSER 22)	25.9	

Notes:

1.   In the process of further validating the information in the WBN Unit 2 FSAR, TVA identified minor administrative/typographical changes to sections previously considered Resolved. TVA addressed these changes to the applicable sections in their submittals and clearly indicated them to the NRC staff. The NRC staff has reviewed and confirmed that the changes made are administrative/typographical and do not impact the NRC staff's conclusions as stated in previous SSERs. Based on this review, no additional review is necessary and this section remains Resolved.

2. During the assessment of the regulatory framework for completion of the project, the NRC staff characterized certain topics as "Open" pending TVA's validation of the information contained in the section. TVA has determined that the information presented in the FSAR remained valid and only identified minor administrative or typographical changes to the section. TVA addressed the changes in their submittals and clearly indicated the changes. The NRC staff reviewed and confirmed that the changes made to the section are administrative/typographical and do not impact its conclusions as stated in previous SSERs. Therefore, no additional review is necessary and the NRC staff considers this section Resolved.

3. In SSER 21, this issue was identified as "Resolved." However, TVA made changes to the Unit 2 FSAR affecting the previous NRC staff conclusions. The NRC staff evaluated the changes and the results are documented in this SSER.

1.8   Confirmatory Issues

At this point in the review, there are some items that have essentially been resolved to the NRC staff's satisfaction, but for which certain confirmatory information has not yet been provided by the applicant. In these instances, the applicant has committed to provide the confirmatory information in the near future. If NRC staff review of this information does not confirm preliminary conclusions on an item, that item will be treated as open, and the NRC staff will report on its resolution in a supplement to this report.

The confirmatory items, with appropriate references to subsections of this report, are noted in Appendix HH.

1.9   License Conditions

The NRC staff proposed two license conditions discussed in Section 2.4.10 of SSER 24.

Flooding Protection Proposed License Condition No. 1:

TVA will submit to the NRC staff by August 31, 2012, for review and approval, a summary of the results of the finite element analysis, which demonstrates that the Cherokee and Douglas dams are fully stable under design basis probable maximum flood loading conditions for the long-term stability analysis, including how the preestablished acceptance criteria were met.

Flooding Protection Proposed License Condition No. 2:

TVA will submit to the NRC staff, before completion of the first operating cycle, its long-term modification plan to raise the height of the embankments associated with the Cherokee, Fort Loudoun, Tellico, and Watts Bar dams. The submittal shall include analyses to demonstrate that, when the modifications are complete, the embankments will meet the applicable structural loading conditions, stability requirements, and functionality considerations to ensure that the design basis probable maximum flood limits are not exceeded at the Watts Bar Nuclear Plant. All modifications to raise the height of the embankments shall be completed within 3 years from the date of issuance of the operating license.

The NRC staff proposed two license conditions discussed in Section 13.6.6.3.22 of SSER 24.

Cyber Security Proposed License Condition 1:

> The licensee shall implement the requirements of 10 CFR 73.54(a)(1)(ii) as they relate to the security computer. Completion of these actions will occur consistent with the full implementation date of September 30, 2014, as established in the licensee's letter dated April 7, 2011, "Response to Request for Additional Information Regarding Watts Bar Nuclear Plant Cyber Security Plan License Amendment Request, Cyber Security Plan Implementation Schedule - Watts Bar Nuclear Plant Unit 1."

Cyber Security Proposed License Condition 2:

> The licensee shall implement the requirements of 10 CFR 73.54(a)(1)(iii) as they relate to the corporate based systems that support emergency preparedness. Completion of these actions will occur consistent with the Watts Bar Nuclear Plant Unit 1 implementation schedule established in the licensee's letter dated April 7, 2011, "Response to Request for Additional Information Regarding Watts Bar Nuclear Plant Cyber Security Plan License Amendment Request, Cyber Security Plan Implementation Schedule - Watts Bar Nuclear Plant Unit 1."

1.10   Unresolved Safety Issues

Section 210 of the Energy Reorganization Act of 1974, as amended, states, in part,

> The Commission shall develop a plan for providing for specification and analysis of unresolved safety issues relating to nuclear reactors and shall take such action as may be necessary to implement corrective measures with respect to such issues.

The NRC staff continuously evaluates the safety requirements used in its review against new information as it becomes available. In some cases, the NRC staff takes immediate action or interim measures to ensure safety. In most cases, however, the initial assessment indicates that immediate licensing actions or changes in licensing criteria are not necessary. In any event, further study may be deemed appropriate to make judgments as to whether existing requirements should be modified. The issues being studied are sometimes called generic safety issues because they are related to a particular class or type of nuclear facility.

The NRC staff documented its original review of Unresolved Safety Issues for WBN Units 1 and 2 in Appendix C to the safety evaluation report (SER; NUREG-0847, June 1982). A discussion of the status of resolution of these generic issues for TVA's application for an operating license for WBN Unit 2 is provided in Appendix C to SSER 23, dated July 2011.

1.13   Implementation of Corrective Action Programs and Special Programs

In 1985, TVA developed a corporate Nuclear Performance Plan (NPP) that identified and proposed corrections to problems concerning the overall management of its nuclear program and a site-specific plan for WBN entitled, "Watts Bar Nuclear Performance Plan." TVA established 18 corrective action programs (CAPs) and 11 special programs (SPs) to address these concerns.

SSER 21, Table 1.13.1 documented the status of NRC staff review of the CAPs and SPs. This SSER and future supplements to the SER, the NRC staff will document its evaluation and closure of open NPP items.

1.13.1 Corrective Action Programs

No.	Title	Program Review Status
(1)	Cable Issues	Resolved
	a. Silicon Rubber Insulated Cable	(See Appendix HH)
	b. Cable Jamming	
	c. Cable Support in Vertical Conduit	
	d. Cable Support in Vertical Trays	
	e. Cable Proximity to Hot Pipes	
	f. Cable Pull-Bys	
	g. Cable Bend Radius	
	h. Cable Splices	
	i. Cable Sidewall Bearing Pressure	
	j. Pulling Cables Through 90° Condulet and Flexible Conduit	
	k. Computer Cable Routing System Software and Database Verification and Validation	
(2)	Cable Tray and Tray Supports	Resolved
(3)	Design Baseline and Verification Program	Resolved
(4)	Electrical Conduit and Conduit Support	Resolved
(5)	Electrical Issues	Resolved
	a. Flexible Conduit Installations	(See Appendix HH)
	b. Physical Cable Separation and Electrical Isolation	
	c. Contact and Coil Rating of Electrical Devices	
	d. Torque Switch and Overload Relay Bypass Capability for Active Safety-Related Valves	
	e. Adhesive-Backed Cable Support Mount	
(6)	Equipment Seismic Qualification	Resolved
(7)	Fire protection	Resolved
(8)	Hanger and Analysis Update Program	Resolved
(9)	Heat Code Traceability	Resolved
(10)	Heating, Ventilation, and Air-Conditioning Duct and Duct Supports	Resolved
(11)	Instrument Lines	Resolved
(12)	Prestart Test Program Plan	Resolved

No.	Title	Program Review Status
(13)	Quality Assurance (QA) Records	Resolved
(14)	Quality-List (Q-List)	Resolved
(15)	Replacement Items Program (Piece Parts)	Resolved
(16)	Seismic Analysis	Resolved
(17)	Vendor Information Program	Resolved
(18)	Welding	Resolved

1.13.2 Special Programs

No.	Title	Program Review Status
(1)	Concrete Quality Program	Resolved
(2)	Containment Cooling	Resolved
(3)	Detailed Control Room Design Review	Resolved
(4)	Environmental Qualifications Program	Resolved
(5)	Master Fuse List	Resolved
(6)	Mechanical Equipment Qualification	Resolved
(7)	Microbiologically Induced Corrosion	Resolved
(8)	Moderate Energy Line Break Flooding	Resolved
(9)	Radiation Monitoring System	Resolved
(11)	Use-As-Is Condition Adverse to Quality	Resolved

1.14   Implementation of Applicable Bulletin and Generic Letter Requirements

From time to time, the NRC staff issues generic requirements or recommendations in the form of orders, bulletins (BLs), generic letters (GLs), regulatory issue summaries, and other documents to address certain safety and regulatory issues. These are generally termed "generic communications."

The table below outlines the status of the resolution of the generic communications. It should be noted that, although many of the generic communications have been documented or otherwise resolved, the NRC staff has determined that there may be circumstances that could result in the need to reopen a previously closed topic.

	Correspondence No.	Title
(1)	GL 1980-14	Light-Water Reactor Primary Coolant System Pressure Isolation Valves.
	TVA Action:	Submit Technical Specifications (TSs) for NRC Review.
	NRC Action:	To be reviewed during validation of TS 3.4.14 submitted February 2, 2010.
(2)	GL 1980-77	Refueling Water Level - Technical Specifications Changes.
	TVA Action:	Submit Technical Specifications for NRC Review.
	NRC Action:	To be reviewed during validation of TS 3.9.5 –TS 3.9.7 submitted February 2, 2010.
(3)	GL 1982-28	Inadequate Core Cooling Instrumentation System.
	TVA Action:	Closed.
	NRC Action:	Closed. Subsumed as part of NRC staff review of Instrumentation and Controls submitted April 8, 2010.
(4)	GL 1983-28	Required Actions Based on Generic Implications of Salem Anticipated Transient without Scram Events (Screened into the Items 4 through 7).
(4.a)	GL 1983-28 (item 3.1)	Post-Maintenance Testing (reactor trip system components).
	TVA Action:	Submit Technical Specifications for NRC Review.
	NRC Action:	To be reviewed during validation of TS Bases 3.0.1 submitted March 4, 2009.

	Correspondence No.	Title
(4.b)	GL 1983-28 (3.2)	Post-Maintenance Testing (All Surveillance Requirement Components).
	TVA Action	Submit Technical Specifications and NRC Review.
	NRC Action	To be reviewed during validation of TS Bases 3.0.1 submitted March 4, 2009.
(4.c)	GL 1983-28 (4.2)	Reactor Trip System Reliability (Preventive Maintenance and Surveillance Program for Reactor Trip Breakers).
	TVA Action	Submit Technical Specifications and NRC Review.
	NRC Action	To be reviewed during NRC staff evaluation of Item 17 of TS Table 3.3.1-1 submitted February 2, 2010.
(4.d)	GL 1983-28 (4.5)	Reactor Trip System Reliability (Automatic Actuation of Shunt Trip Attachment).
	TVA Action	Submit Technical Specifications and NRC Review.
	NRC Action	To be reviewed during NRC staff evaluation of Item 18 of TS Table 3.3.1-1 submitted February 2, 2010.
(8)	GL 1986-09	Technical Resolution of Generic Issue B-59, (N-1) Loop Operation in BWRs and PWRs.
	TVA Action	Submit Technical Specifications for NRC Review.
	NRC Action	To be reviewed during validation of TS 3.4.4 - TS 3.4.8 submitted February 2, 2010.
(9)	GL 1988-20	Individual Plant Examination for Severe Accident Vulnerability.
	TVA Action	Closed.
	NRC Action	Closed. NRC letter dated August 12, 2011 (ADAMS Accession No. ML111960228).
(10)	GL 1988-20s1	Initiation of the Individual Plant Examination for Severe Accident Vulnerabilities — 10 CFR 50.54.
	TVA Action	Closed.
	NRC Action	Closed. NRC letter dated August 12, 2011 (ADAMS Accession No. ML111960228).

	Correspondence No.	Title
(11)	GL 1988-20s2	Individual Plant Examination for Severe Accident Vulnerability. Accident Management Strategies for Consideration in the Individual Plant Examination Process.
	TVA Action	Closed.
	NRC Action	Closed. NRC letter dated August 12, 2011 (ADAMS Accession No. ML111960228).
(12)	GL 1988-20s3	Individual Plant Examination for Severe Accident Vulnerability. Completion of Containment Performance Improvement Program and Forwarding of Insights for Use in the IPE for Severe Accident Vulnerabilities.
	TVA Action	Closed.
	NRC Action	Closed. NRC letter dated August 12, 2011 (ADAMS Accession No. ML111960228).
(13)	GL 1988-20s4	Individual Plant Examination of External Events (IPEEE) for Severe Accident Vulnerabilities.
	TVA Action	Closed.
	NRC Action	Closed. NRC letter dated September 20, 2011 (ADAMS Accession No. ML111960300).
(14)	GL 1988-20s5	Individual Plant Examination of External Events (IPEEE) for Severe Accident Vulnerabilities - 10 CFR 50.54(f).
	TVA Action	Closed.
	NRC Action	Closed. NRC letter dated September 20, 2011 (ADAMS Accession No. ML111960300).
(15)	GL 1989-04	Guidelines on Developing Acceptable Inservice Testing Programs.
	TVA Action	The proposed approach has been approved for WBN Unit 1; the same approach will be proposed for use on WBN Unit 2 without change.
	NRC Action	Open.

	Correspondence No.	Title
(16)	GL 1989-21	Request for Information Concerning Status of Implementation of Unresolved Safety Issue Requirements.
	TVA Action	TVA provided an updated status of unresolved safety issues on September 26, 2008, as supplemented on December 2, 2010, and January 25, 2011.
	NRC Action	Closed. See Appendix C of SSER 23.
(17)	GL 1990-06	Resolution of Generic Issues 70, "PORV [power-operated relief valve] and Block Valve Reliability," and 94, "Additional LTOP [low-temperature overpressure] Protection for PWRs."
	TVA Action	Submit Technical Specifications for NRC Review.
	NRC Action	To be reviewed during validation of TS 3.4.11 - TS 3.4.12 submitted February 2, 2010.
(18)	GL 1992-08	Thermo-Lag 330-1 Fire Barriers.
	TVA Action	The proposed approach has been approved for WBN Unit 1; the same approach will be proposed for use on WBN Unit 2 without change.
	NRC Action	Open. Pending NRC staff inspection verification.
(19)	GL 1995-03	Circumferential cracking of Steam Generator (SG) Tubes.
	TVA Action	The proposed approach has been approved for WBN Unit 1; the same approach was submitted for use on WBN Unit 2 without change.
	NRC Action	Closed. NRC Letter dated January 21, 2010 (ADAMS Accession No. ML093631061).
(20)	GL 1995-05	Voltage –Based Repair Criteria for Westinghouse Steam Generator Tubes affected by Outside Diameter Stress Corrosion Cracking.
	TVA Action	The proposed approach has been approved for WBN Unit 1; the same approach was submitted for use on WBN Unit 2 without change.
	NRC Action	Closed. NRC Letter dated January 21, 2010 (ADAMS Accession No. ML093631061).

Correspondence No.	Title
(21) GL 1996-06	Assurance of Equipment Operability and Containment Integrity During Design-Basis Accident Conditions.
TVA Action	The proposed approach has been approved for WBN Unit 1; the same approach will be proposed for use on WBN Unit 2 without change.
NRC Action	Closed. NRC Letter dated January 21, 2010 (ADAMS Accession No. ML100130227).
(22) GL 1995-07	Pressure Locking and Thermal Binding of Safety-Related Power-Operated Gate Valves (Not identified in SSER 21 as "Open").
TVA Action	The proposed approach has been approved for WBN Unit 1; the same approach will be proposed for use on WBN Unit 2 without change.
NRC Action	Closed. NRC letter dated August 12, 2010 (ADAMS Accession No. ML100190443).
(23) GL 1997-01	Degradation of Control Rod Drive Mechanism Nozzle and Other Vessel Closure Head Penetrations.
TVA Action	The proposed approach has been approved for WBN Unit 1; the same approach will be proposed for use on WBN Unit 2 without change.
NRC Action	Closed. NRC Letter dated June 30, 2010 (ADAMS Accession No. ML100539515).
(24) GL 1997-04	Assurance of Sufficient Net Positive Suction Head for Emergency Core Cooling and Containment Heat Removal Pumps Integrity During Design-Basis Accident Conditions.
TVA Action	The proposed approach has been approved for WBN Unit 1; the same approach was submitted for use on WBN Unit 2 without change.
NRC Action	Closed. NRC Letter dated February 18, 2010 (ADAMS Accession No. ML100200375).

	Correspondence No.	Title
(25)	GL 1997-05	SG Tube Inspection Techniques.
	TVA Action	The proposed approach has been approved for WBN Unit 1; the same approach was submitted for use on WBN Unit 2 without change.
	NRC Action	Closed. NRC Letter dated January 21, 2010 (ADAMS Accession No. ML093631061).
(26)	GL 1997-06	Degradation of SG Internals.
	TVA Action	The proposed approach has been approved for WBN Unit 1; the same approach was submitted for use on WBN Unit 2 without change.
	NRC Action	Closed. NRC Letter dated January 21, 2010 (ADAMS Accession No. ML093631061).
(27)	GL 1998-02	Loss of Reactor Coolant Inventory and Associated Potential for Loss of Emergency Mitigation Functions While in a Shutdown Condition.
	TVA Action	The proposed approach has been approved for WBN Unit 1; the same approach will be proposed for use on WBN Unit 2 without change.
	NRC Action	Closed. NRC Letter dated May 11, 2010 (ADAMS Accession No. ML101200155).
(28)	GL 1998-04	Potential for Degradation of the ECCS [Emergency Core Cooling System] and the Containment Spray System after a LOCA because of Construction and Protective Coating Deficiencies and Foreign Material in Containment.
	TVA Action	The proposed approach has been approved for WBN Unit 1; the same approach was submitted for use on WBN Unit 2 without change.
	NRC Action	Closed. NRC Letter dated February 1, 2010 (ADAMS Accession No. ML100260594).

	Correspondence No.	Title
(29)	GL 2003-01	Control Room Habitability.
	TVA Action	No action or documentation is provided to show the NRC staff has reviewed the item for WBN Unit 2, and the resolution is through submittal of a technical specification.
	NRC Action	Closed. NRC Letter dated February 1, 2010 (ADAMS Accession No. ML100270076).
(30)	GL 2004-01	Requirements for SG Tube Inspection.
	TVA Action	The proposed approach has been approved for WBN Unit 1; the same approach was submitted for use on WBN Unit 2 without change.
	NRC Action	Closed. NRC Letter dated January 21, 2010 (ADAMS Accession No. ML093631061).
(31)	GL 2004-02	Potential Impact of Debris Blockage on Emergency Recirculation during Design-Basis Accidents at PWRs.
	TVA Action	The proposed approach has been approved for WBN Unit 1; the same approach was submitted for use on WBN Unit 2 without change.
	NRC Action	Open.
(32)	GL 2006-01	SG Tube Integrity and Associated Technical Specifications.
	TVA Action	No action or documentation is provided to show the NRC staff has reviewed the item for WBN Unit 2, and the resolution is through submittal of a technical specification.
	NRC Action	Closed. NRC Letter dated January 21, 2010 (ADAMS Accession No. ML093631061) (See Appendix HH).
(33)	GL 2006-02	Grid Reliability and the Impact on Plant Risk and the Operability of Offsite Power.
	TVA Action	The proposed approach has been approved for WBN Unit 1; the same approach was submitted for use on WBN Unit 2 without change.
	NRC Action	Closed. NRC Letter dated January 21, 2010 (ADAMS Accession No. ML093631061) (See Appendix HH).

Correspondence No.	Title
(34) GL 2006-03	Potentially Nonconforming Hemyc and MT Fire Barrier Configurations.
TVA Action	The proposed approach has been approved for WBN Unit 1; the same approach was submitted for use on WBN Unit 2 without change.
NRC Action	Closed. NRC Letter February 25, 2010 (ADAMS Accession No. ML100470398).
(35) GL 2007-01	Inaccessible or Underground Power Cable Failures that Disable Accident Mitigation Systems or Cause Plant Transients.
TVA Action	The proposed approach has been approved for WBN Unit 1; the same approach was submitted for use on WBN Unit 2 without change.
NRC Action	Closed. NRC Letter dated January 26, 2010 (ADAMS Accession No. ML100120052).
(36) GL 2008-01	Managing Gas Accumulation in Emergency Core Cooling, Decay Heat Removal, and Containment Spray Systems.
TVA Action	TVA submitted the information requested by the GL.
NRC Action	Closed. NRC letter dated August 23, 2011 (ADAMS Accession No. ML112232205).
(37) BL 1992-01 and Supplement 1	Failure of Thermo-Lag 330 Fire Barrier System to Perform its Specified Fire Endurance Function.
TVA Action	The proposed approach has been approved for WBN Unit 1; the same approach will be proposed for use on WBN Unit 2 without change.
NRC Action	Open. Pending NRC staff inspection verification.
(38) BL 1996-01	Control Rod Insertion Problems (PWR)
TVA Action	The proposed approach has been approved for WBN Unit 1; the same approach was submitted for use on WBN Unit 2 without change.
NRC Action	Closed. NRC letter dated May 3, 2010 (ADAMS Accession No. ML101200035) required Confirmatory Action (See Appendix HH).

	Correspondence No.	Title
(39)	BL 1996-02	Movement of Heavy Loads Over Spent Fuel, Over Fuel In the Reactor Core, or Over Safety-Related Equipment.
		The proposed approach has been approved for WBN Unit 1; the same approach was submitted for use on WBN Unit 2 without change.
		Closed. NRC Letter dated March 4, 2010 (ADAMS Accession No. ML100480062).
(40)	BL 2001-01	Circumferential Cracking of Reactor Pressure Vessel (RPV) Head Penetration Nozzles.
	TVA Action	The proposed approach has been approved for WBN Unit 1; the same approach was submitted for use on WBN Unit 2 without change.
	NRC Action	Closed. See NRC Letter dated June 30, 2010 (ADAMS Accession No. ML 100539515).
(41)	BL 2002-01	RPV Head Degradation and Reactor Coolant Pressure Boundary Integrity.
	TVA Action	The proposed approach has been approved for WBN Unit 1; the same approach was submitted for use on WBN Unit 2 without change.
	NRC Action	Closed. See NRC Letter dated June 30, 2010 (ADAMS Accession No. ML 100539515).
(42)	BL 2002-02	RPV Head and Vessel Head Penetration Nozzle Inspection Program.
	TVA Action	The proposed approach has been approved for WBN Unit 1; the same approach was submitted for use on WBN Unit 2 without change.
	NRC Action	Closed. See NRC Letter dated June 30, 2010 (ADAMS Accession No. ML100539515).

Correspondence No.	Title
(43) BL 2003-02	Leakage from RPV Lower Head Penetrations and Reactor Coolant Pressure Boundary Integrity.
TVA Action	The proposed approach has been approved for WBN Unit 1; the same approach was submitted for use on WBN Unit 2 without change.
NRC Action	Closed. NRC Letter dated January 21, 2010 (ADAMS Accession No. ML093631061).
(44) BL 2004-01	Inspection of Alloy 82/182/600 Materials Used in the Fabrication of Pressurizer Penetrations and Steam Space Piping Connections at PWRs.
TVA Action	The proposed approach has been approved for WBN Unit 1; the same approach was submitted for use on WBN Unit 2 without change.
NRC Action	Closed. NRC letter dated August 4, 2010 (ADAMS Accession No. ML102080017).
(45) BL 2007-01	Security Officer Attentiveness.
TVA Action	The proposed approach has been approved for WBN Unit 1; the same approach will be proposed for use on WBN Unit 2 without change.
NRC Action	Closed. NRC letter dated March 25, 2010 (ADAMS Accession No. ML100770549).
(46) BL 20011-01	Mitigating Strategies
TVA Action	The proposed approach has been approved for WBN Unit 1; an updated approach will be proposed for use on WBN Unit 2 without change.
NRC Action	Open.
(47) BL 2012-01	Design Vulnerability In Electric Power System
TVA Action	The proposed approach has submitted for WBN Unit 1 and Unit 2 and is still under review by the NRC staff.
NRC Action	Open.

NUREG-0737, TMI Action Items (TVA letter dated September 14, 1981, applies to all of the following NUREG-0737 issues):

	Correspondence No.	Title
(48)	NUREG-0737 Item I.B.1.2	Independent Safety Engineering Group.
	TVA Action	The proposed approach has been approved for WBN Unit 1; the same approach will be proposed for use on WBN Unit 2 without change.
	NRC Action	Open.
(49)	NUREG-0737 Item I.D.1	Control Room Design Review (CRDR).
	TVA Action	The proposed approach has been approved for WBN Unit 1; the same approach will be proposed for use on WBN Unit 2 without change.
	NRC Action	Closed in SSER 22, Section 18.2.
(50)	NUREG-0737 Item II.B.3	Post-accident Sampling.
	TVA Action	No action or documentation is provided to show the NRC staff has reviewed the item for WBN Unit 2, and the resolution is through submittal of a technical specification.
	NRC Action	Closed in SSER 24, Section 9.3.2.
(51)	NUREG-0737 Item II.E.4.2	Containment Isolation Dependability.
	TVA Action	No action or documentation is provided to show the NRC staff has reviewed the item for WBN Unit 2, and the resolution is through submittal of a technical specification.
	NRC Action	Open.
(52)	NUREG-0737 Item II.F.2	Instrumentation for Detection of Inadequate Core-Cooling.
	TVA Action	Open.
	NRC Action	Open. See SSER 23, Section 4.4.8.

Correspondence No.	Title
(53) NUREG-0737 Item II.K.3.3	Reporting SV/RV Failures/Challenges.
TVA Action	No action or documentation is provided to show the NRC staff has reviewed the item for WBN Unit 2, and the resolution is through submittal of a technical specification.
NRC Action	Closed in SSER 22, Section 13.5.3.
(54) NUREG-0737 Item II.K.3.10	Anticipatory Trip at High Power.
TVA Action	No action or documentation is provided to show the NRC staff has reviewed the item for WBN Unit 2, and the resolution is through submittal of a technical specification.
NRC Action	Open.
(55) NUREG-0737 Item III.D.1.1	Primary Coolant Outside Containment.
TVA Action	No action or documentation is provided to show the NRC staff has reviewed the item for WBN Unit 2, and the resolution is through submittal of a technical specification.
NRC Action	Open.
(56) NUREG-0737 Item III.D.3.4	Control-Room Habitability.
TVA Action	The proposed approach has been approved for WBN Unit 1; the same approach will be proposed for use on WBN Unit 2 without change.
NRC Action	Closed in SSER 22, Section 6.4.
(57) IEB 75-08	PWR Pressure Instrumentation.
TVA Action	The item has been approved either for both units at WBN or explicitly for WBN Unit 2; however, a change to the original approval requires submittal of the technical specifications and NRC staff review.
NRC Action	Open.

	Correspondence No.	Title
(58)	IEB 77-04	Calculation Error Affecting Performance of a System for Controlling pH of Containment Sump Water Following a LOCA.
	TVA Action	The item has been approved either for both units at WBN or explicitly for WBN Unit 2; however, a change to the original approval requires submittal of the technical specifications and NRC staff review.
	NRC Action	Open.

2   SITE CHARACTERISTICS

2.3   Meteorology

Disposition of Open Items (Appendix HH)

2.3.3   Onsite Meteorological Measurements Program

Open Item 136

Open Item 136 states the following:

> The JFD [joint frequency distribution] summary for the data from 1991 through 2010 provided by letter dated November 7, 2011, and a discussion of the long-term representativeness of these data should be provided in the WBN Unit 2 FSAR. Upon receipt of the updated FSAR, the NRC staff will confirm that these updates have been made by TVA.

TVA updated the FSAR in Amendment 107 to the FSAR to to address the open item. The NRC staff verified the update to the FSAR. Therefore, **Open Item 136 is closed**.

2.3.4   Short-Term (Accident) Atmospheric Dispersion Estimates

Open Item 137

Open Item 137 states the following:

> The NRC staff will confirm, upon receipt, that TVA integrated the updated CR χ/Q values from its letter dated September 15, 2011, into a future amendment of the FSAR.

The NRC staff reviewed Amendment 107 to the FSAR and verified that it included the updated values. Therefore, **Open Item 137 is closed**.

Open Item 138

Open item 138 states the following:

> Upon receipt of the updated ODCM [Offsite Dose Calculation Manual], the NRC staff will confirm that corresponding revisions related to the updated annual average χ/Q and D/Q values have been made to the ODCM.

The NRC staff reviewed the updated ODCM and Amendment 109 to the FSAR, and verified that the updated values had been included. Therefore, **Open Item 138 is closed**.

3   DESIGN CRITERIA

3.9 Mechanical Systems and Components

Disposition of Open Items (Appendix HH)

3.9.5 Reactor Pressure Vessel Internals

Open Item 71

By letter dated March 11, 2010 (Agencywide Documents Access and Management System (ADAMS) Accession No. ML100550007), the NRC staff asked TVA to provide information on the type of heat treatment of nickel-based Alloy X-750 components of the reactor vessel internals (RVIs) at WBN Unit 2, and to provide information as to whether the plant should replace these components before the start of operation if the Alloy X-750 components are fabricated from material with lower stress corrosion cracking (SCC) resistance. Alloy X-750 components under the heat treatment condition designated as high-temperature heat treated (HTH[1]) are generally known to have better resistance to SCC than Alloy X-750 components under other heat treatment conditions do. By letter dated April 9, 2010 (ADAMS Accession No. ML101040573), TVA partially responded by indicating that the heat treatment of the clevis insert bolts was the BH[2] [hot worked and aged] condition. By letter dated June 28, 2010 (ADAMS Accession No. ML101790399), TVA committed to replace, before WBN Unit 2 operation, the current clevis insert bolts used in WBN Unit 2 with HTH-treated Alloy X-750 bolts of the latest design that have a rolled thread and a larger radius on the undercut of the cap screw head. By letter dated April 21, 2011 (ADAMS Accession No. ML111110513), TVA withdrew the previous commitment and provided information supporting the commitment withdrawal. **Open Item 71** (Appendix HH) states that, "...TVA withdrew its commitment to replace the Unit 2 clevis insert bolts. TVA should provide further justification for the decision to not replace the bolts to the NRC staff." TVA provided additional information supporting the withdrawal of the commitment by letter dated October 24, 2011 (ADAMS Accession No. ML11299A130).

The NRC staff reviewed TVA's justification for withdrawal of the commitment to replace the Alloy X-750 clevis insert bolts. The following summary lists the NRC staff's understanding of TVA's justification for retaining the existing bolts:

- Only one documented failure of clevis insert bolts has occured, this failure was discovered at Donald C. Cook Nuclear Plant, Unit 1 (D.C. Cook, Unit 1) in 2010. The root cause of this failure remains indeterminate.

- The D.C. Cook, Unit 1 failure has not been definitively attributed to SCC.

- Clevis insert bolt failures would be detected through the normal American Society of Mechanical Engineers Boiler & Pressure Vessel Code (ASME Code), Section XI in service inspections conducted during each 10-year inspection interval.

---

[1] The HTH condition is hot worked, annealed at 2000 °F (1093 °C), and aged at 1300 °F (704 °C) for 20 hours, then air cooled (Mills, W. J, Lebo, M. R., and Kearns, J. J., "Hydrogen Embrittlement, Grain Boundary Segregation, and Stress Corrosion Cracking of Alloy X-750 in Low-and High-Temperature Water," in Metallurgical and Materials Transactions A Volume 30, Number 6, 1579-1596, DOI: 10.1007/s11661-999-0095-8).

[2] The BH condition is hot worked, then aged at 1300 °F (704 °C) for 20 hours, then air cooled.

- Industry experience from inservice inspections does not support the existence of a widespread problem with clevis insert bolt failures. In its letter dated October 24, 2011, TVA cited information from eight plants indicating that inservice inspections within the last 5 years found no evidence of clevis insert bolt failures, and cited 12 more plants that should have performed inservice inspections within this time period with no reports of clevis insert bolt failures. In its previous letter dated April 21, 2011, TVA cited approximately 15 years of operating experience at WBN Unit 1, with no evidence of degradation of the clevis insert bolts.

The NRC staff reviewed TVA's justification for withdrawal of the commitment, as summarized above. The NRC staff agrees with TVA's assessment of operating experience related to clevis insert bolts, however, it notes that other Alloy X-750 components in RVI have experienced failures caused by SCC, most notably control rod guide tube support pins (split pins) in RVIs designed by Westinghouse. Alloy X-750 baffle-to-former bolts in a German PWR (Biblis-A) also experienced cracking attributed to SCC. Most of the failed split pins and the baffle-to-former bolts were in the AH[3] or a similar heat treatment condition; however, even some split pins in the HTH condition experienced SCC, which was attributed to poor surface condition. Furthermore, information from the Electric Power Research Institute Materials Reliability Program indicates that SSC is the most likely cause of the D.C. Cook, Unit 1 clevis insert bolt failure. Therefore, the NRC staff recommends that TVA closely monitor operating experience related to RVI components fabricated from Alloy X-750, including both clevis insert bolts and other components, to determine whether it will need to replace the clevis insert bolts in the future.

Based on the foregoing, the NRC staff concludes that TVA may withdraw its commitment to replace the existing clevis insert bolts before plant startup. Therefore, **Open Item 71 is closed**. The NRC staff recommends that TVA monitor relevant industry experience and that it replace the bolts in the future, if necessary, based on operating experience.

---

[3] The AH condition is hot rolled, and equalized at 1625 °F (885 °C) followed by 20 hours at 1300 °F (704 °C).

6   ENGINEERED SAFETY FEATURES

Disposition of Open Items (Appendix HH)

6.2   Containment Systems

6.2.6   Containment Leakage Testing

Open Item 47

Open Item 47 states the following:

> The NRC staff noted that TVA's changes to Section 6.2.6 in FSAR
> Amendment 97, regarding the implementation of Option B of Appendix J, were
> incomplete, because several statements remained regarding performing
> water-sealed valve leakage tests "as specified in 10 CFR [Part] 50, Appendix J."
> With the adoption of Option B, the specified testing requirements are no longer
> applicable; Option A to Appendix J retains these requirements. The NRC
> discussed this discrepancy with TVA in a telephone conference on
> September 28, 2010. TVA stated that it would remove the inaccurate reference
> to Appendix J for specific water testing requirements in a future FSAR
> amendment.

TVA updated the FSAR in Amendments 104 and 105 to correct the discrepancy. The NRC staff
verified the correction. Therefore, **Open Item 47 is closed**.

6.5   Engineered Safety Feature Filter Systems

6.5.3   Fission Product Control System and Structures

Open Item 48

Open Item 48 states the following:

> The NRC staff should verify that its conclusions in the review of FSAR
> Section 15.4.1 do not affect the conclusions of the staff regarding the
> acceptability of Section 6.5.3.

The NRC staff reviewed FSAR Section 15.4.1 in Supplemental Safety Evaluation Report 25 and
concluded that it is acceptable. Therefore, **Open Item 48 is closed**.

7   INSTRUMENTATION AND CONTROLS

7.2   Reactor Trip System

7.2.1   System Description

7.2.1.1   Eagle 21 System

Disposition of Open Items (Appendix HH)

Open Item 65

Open Item 65 states, "TVA should provide justification to the staff regarding why different revisions of WCAP-13869 are referenced in WBN Unit 1 and Unit 2." WBN Unit 1 FSAR sections 7.1 and 7.2 referenced WCAP-13869, "Reactor Protection System Diversity in Westinghouse Pressurized Water Reactors," Revision 1, while the WBN Unit 2 FSAR referenced Revision 2.

In its letter dated October 29, 2010 (ADAMS Accession No. ML103120711), Item 36, TVA provided a section entitled "WCAP-13869 Revision 1 to Revision 2 Change Analysis." TVA stated that the design bases for the response to a feedwater break inside containment accident, as documented in Chapter 15 of the WBN Unit 2 FSAR, is the same as that for Unit 1. In its letter dated March 31, 2011 (ADAMS Accession No. ML110950331), Item 14, TVA stated that, because WBN Unit 2 is required to match the Unit 1 licensing basis to the extent practical, it would revise the WBN Unit 2 FSAR to agree with the WBN Unit 1 FSAR, which references Revision 1.

The NRC staff reviewed Sections 7.1 and 7.2 of WBN Unit 2 FSAR Amendment 104, submitted on June 3, 2011 (ADAMS Accession No. ML111780527), and verified that WCAP-13869, Revision 1, is referenced instead of Revision 2. Therefore, **Open Item 65 is closed**.

7.5   Safety-Related Display Instrumentation

7.5.2   Post-Accident Monitoring System

7.5.2.2   Common Qualified Platform - Postaccident Monitoring System

Disposition of Open Items (Appendix HH)

Open Item 94

Open Item 94 states, "TVA should provide to the staff either information that demonstrates that the WBN Unit 2 Common Q PAMS [postaccident monitoring system] meets the applicable requirements in Institute of Electrical and Electronics Engineers (IEEE) Std. 603-1991, or justification for why the Common Q PAMS should not meet those requirements."

By letter dated November 18, 2011 (ADAMS Accession No. ML113130218), the NRC staff requested additional information regarding this open item. By letter dated March 9, 2012 (ADAMS Accession No. ML12073A391), TVA responded (See Letter Enclosure 1, Item No. 1 and Attachment 4). In Attachment 4, TVA documented the evaluation of conformance with each requirement of Clauses 5 and 6 (Clause 4 is addressed under Open Item 105). TVA stated that the WBN Unit 2 Common Q PAMS does not automatically perform any protection or control functions; therefore, the clauses applicable to these functions are not applicable to the WBN Unit 2 Common Q PAMS. The NRC staff reviewed TVA's explanation and agreed that IEEE Std. 603-1991 Clauses 5.2, 5.8.3, 6.1, 6.2.1, 6.3, 6.6, 6.7, and 6.8 are not applicable. The NRC staff confirmed TVA's explanations for the remaining clauses by a review of the design documentation for the WBN Unit 2 Common Q PAMS; therefore, the NRC staff concludes that the WBN Unit 2 Common Q PAMS meets the regulatory requirements in these clauses. **Open Item 94 is closed**.

Open Item 101

Open Item 101 states that, "TVA should demonstrate that the WBN Unit 2 Common Q PAMS application software is in conformance with RG 1.168, Revision 1 or provide justification for not conforming."

In its letter dated December 22, 2011 (ADAMS Accession No. ML12018A213; letter Item 9 and Attachment 9), TVA provided the following regulatory commitment to revise the FSAR as follows:

> The changes to FSAR Table 7.1-1 for RG 1.168, IEEE 1012 and IEEE 1028 shown in Attachment 9 will be incorporated into Amendment 108 of the FSAR. (See Letter Item 9)

TVA's commitment to update the FSAR to identify that the Common Q PAMS is required to be in conformance with RG 1.168, "Verification, Validation, Reviews, and Audits for Digital Computer Software Used in Safety Systems of Nuclear Power Plants," Revision 0, issued September 1997; IEEE Std. 1012-1986, "IEEE Standard for Software Verification and Validation Plans"; and IEEE Std. 1028-1988, "IEEE Standard for Software Reviews and Audits," is acceptable to the NRC staff, because the Common Q PAMS was developed in accordance with the NRC-approved "Software Program Manual for Common Q Systems" (ADAMS Accession No. ML050350234). Therefore, **Open Item 101 is closed**.

Open Item 105

Open Item 105 states, "TVA should provide to the NRC staff an acceptable description of how the WBN Unit 2 Common Q PAMS SysRS and SRS implement the design-basis requirements of IEEE Std. 603-1991, Clause 4."

By letter dated November 18, 2011 (ADAMS Accession No. ML113130218), the NRC staff requested additional information regarding this open item. By letter dated March 9, 2012 (ADAMS Accession No. ML12073A391), TVA responded (See Letter Enclosure 1, Item No. 5 and Attachment 4).

The WBN Unit 2 Common Q PAMs is functionally equivalent to the Unit 1 Inadequate Core Cooling Monitor (ICCM-86); therefore, the WBN Unit 2 design uses the Unit 1 ICCM-86 specification as a basis. In addition, the placement of displays and associated controls for Unit 2 was subjected to a human factors engineering review during the design process for EDCR 52351, "Common Q PAMS and the Control Room Design Review (CRDR)." The NRC staff did not intend for its technical review to be a re-evaluation of these aspects; however, compliance of the WBN Unit 2 design with these aspects, as well as the design basis events, was addressed. Because the ICCM-86 specification was part of the contractual requirements, assessment of compliance was addressed primarily through the assessment of the verification and validation (V&V) activities. However, the design bases events, were not explicitly addressed in the V&V documentation; therefore the NRC staff used Open Item 105 to explicitly address documentation with respect to the design basis events.

To address Clause 4 of IEEE Std. 603-1991(i.e., the adequacy of the design basis event documentation), TVA evaluated the emergency operating instructions (EOIs) and the abnormal operating procedures (AOPs) to ensure that the CQ PAMS equipment could fulfill the requirements of these documents. In addition, TVA mapped the limiting design basis events (i.e., those documented in Chapter 15 of the FSAR) to the EOIs and AOPs to ensure that all of the events were properly addressed. Together, these two analyses demonstrate that the WBN Unit 2 Common Q PAMS has the appropriate features and performance to address the documented design basis events.

Clause 4 of IEEE 603-1991 requires that the design bases event documentation include:

- the identification of applicable modes of operation of the generating station, the design basis event applicable to each mode, the initial conditions for each event and the allowable limits of plant conditions for each event.
- the safety functions and corresponding protective actions for each design basis event.
- the variables that must be monitored for each manually controlled protective action, and the associated analytical limits, ranges, and rates of change of these variables.
- the following minimum criteria for each protective action whose operation may be controlled by manual means initially or subsequent to initiation:
  - the points in time and the plant conditions during which manual control is allowed.
  - the justification for permitting initiation or control subsequent to initiation solely by manual means.
  - the variables that shall be displayed for the operator to use in taking manual action.

TVA's analysis demonstrates that the WBN Unit 2 Common Q PAMS addresses the events described in the design basis by tracing the limiting design basis events to the EOIs and AOPs; however, the mapping is not a one-to-one mapping. The EOIs and AOPs are symptom-based rather than event-based. In some cases, more than one event may be addressed by an EOI or AOP, but in other cases more than one EOI or AOP is required to address a single event.

The TVA analysis also demonstrates that Common Q PAMS meets the requirements of the EOIs and AOPs by identifying the requirements in the Common Q PAMS specifications that envelop the performance assumed by the EOIs and AOPs. Based on these analyses, the NRC staff concludes that the design basis documentation for the Common Q PAMS includes the material required by Clause 4 of IEEE 603-1991, and that the Common Q PAMS implements the associated requirements; therefore, the NRC staff concludes that the Common Q PAMS conforms to the regulatory requirements. **Open Item 105 is closed**.

Open Item 108

Open Item 108 states, "TVA should demonstrate to the NRC staff that there are no synergistic effects between temperature and humidity for the Common Q PAMS equipment."

TVA provided a response to the item by letter dated September 1, 2011 (ADAMS Accession No. ML112440576; Enclosure 1, Item 4), as supplemented by letters dated December 22, 2011 (ADAMS Accession No. ML12018A213; Enclosure 1, Item 4, and

Attachments 3 and 4), and January 19, 2012 (ADAMS Accession No. ML12023A191; Item 1 and Attachment 1). The NRC staff reviewed the calculations and analyses provided by TVA and confirmed that the qualification environment includes both temperature and humidity that synergistically bounds the worst case environment in which the equipment is required to operate. Therefore, **Open Item 108 is closed**.

Open Item 110

Open Item 110 states, "TVA should provide information to the NRC staff describing how the WBN Unit 2 Common Q PAMS design supports periodic testing of the reactor vessel level indicating system (RVLIS) function." TVA responded to the item by letter dated September 1, 2011 (ADAMS Accession No. ML11257A048; Enclosure 1, Item 5).

TVA stated that the Common Q PAMS includes a common set of tests to confirm that the equipment is working properly (e.g., annunciator test, analog output test, and input/output simulator connection). The Common Q PAMS also includes self-checks to ensure that the software has not changed. In addition to these built-in capabilities, the RVILIS can be tested by the normal channel calibration methodologies. Based on these capabilities, the NRC staff finds that the Common Q PAMS RVILIS meets the requirements in Clause 5.7, "Capability for Test and Calibration"of IEEE Std. 603-1991. Therefore, **Open Item 110 is closed**.

Open Item 111

Open Item 111 states, "TVA should confirm to the staff that there are no changes required to the technical specifications as a result of the modification installing the Common Q PAMS. If any changes to the technical specifications are required, TVA should provide the changes to the NRC staff for review." The NRC staff reviewed the Unit 2 technical specifications (TSs), Developmental Revision F (Enclosure 1 to TVA's letter dated August 10, 2011; ADAMS Accession No. ML11228A046), and concluded that changes from the Unit 1 TSs are acceptable, because they comply with 10 CFR 50.36. Therefore, **Open Item 111 is closed**.

7.5.2.2.3 Technical Evaluation

Open Item 98

INTRODUCTION

On July 21, 2011, the NRC published NUREG-0847, "Safety Evaluation Report Related to the Operation of Watts Bar Nuclear Plant, Unit 2," Supplement 23 (SSER 23; ADAMS Accession No. ML11206A499). SSER 23, Section 7.5.2.2, "Common Qualified Platform-Post Accident Monitoring System," contained several open items. Open Item 98 was identified in Section 7.5.2.2.3.12 and stated "TVA should demonstrate that the WBN Unit 2 Common Q PAMS is in conformance with RG 1.152, Revision 2, or provide justification for not conforming." Since the original submission of the licensing action, RG 1.152 "Criteria for Use of Computers in Safety Systems of Nuclear Power Plants," has been updated to Revision 3 to better align with current regulations, and TVA has indicated that it intends for WBN Unit 2 to address the guidance contained in Revision 3.

To support closure of this open item, the licensee submitted a letter on September 1, 2011 (ADAMS Accession No. ML11257A048), with attachments containing WCAP-17427-P, Revision 1 (Attachment 6, ADAMS Accession No. ML11257A061) and TVA document

"Common Q PAMS Secure Operational Environment per Regulatory Guide 1.152 Revision 3" (Attachment 9, ADAMS Accession No. ML11257A050), to demonstrate how they meet the guidance contained in RG 1.152, Revision 3. The NRC staff transmitted a request for additional information dated November 17, 2011 (ADAMS Accession No. ML11314A110). The licensee responded to the various questions in letters dated December 22, 2011 (ADAMS Accession No. ML12018A213), March 2, 2012 (ADAMS Accession No. ML12073A224) and March 9, 2012 (ADAMS Accession No. ML120730655).

Although some of the controls, processes, and features that are applied to achieve a Secure Development and Operational Environment (SDOE) will also serve a cyber security function, this safety evaluation section only assesses the system's compliance with 10 CFR Part 50 licensing requirements. The evaluation of TVA's adherence to 10 CFR 73.54 "Protection of Digital Computer and Communication Systems and Networks," and protection of the licensee's digital assets from cyber threats is addressed under other regulatory programs specifically designed for that purpose.

REGULATORY EVALUATION

The following regulations and guidance documents pertain to the establishment of a secure development and operational environment for the WBN Unit 2 PAMS.

- The regulation at 10 CFR 50.55a(h) requires compliance with IEEE Std. 603-1991 and the correction sheet dated January 30, 1995. IEEE Std. 603-1991 contains the following clauses:
  - Clause 5.6.3 addresses independence between safety systems and other systems.
  - Clause 5.9 addresses control of access.

- General Design Criterion (GDC) 21, "Protection system reliability and testability," of Appendix A, "General Design Criteria for Nuclear Power Plants," to 10 CFR Part 50 requires that the protection system be designed for high functional reliability and inservice testability commensurate with the safety functions to be performed.

- Criterion III, "Design Control," of Appendix B, "Quality Assurance Criteria for Nuclear Power Plants and Fuel Reprocessing Plants," to 10 CFR Part 50 requires, in part, that licensees specify appropriate quality standards and provide design control measures for verifying or checking the adequacy of safety system designs.

Revision 3 to RG 1.152 contains five regulatory positions regarding establishment of an SDOE to ensure the high functional reliability of digital safety systems to support compliance with the above cited regulations.

TECHNICAL EVALUATION

The information below is the NRC staff's evaluation of the WBN Unit 2 PAMS against the provisions of Revision 3 to RG 1.152. For the secure development environment evaluation, measures taken to ensure the reliability of the Common Q platform software and the PAMS application were evaluated separately. For the secure operational environment, an assessment was performed against the measures taken to ensure the reliability of the PAMS by protecting it from both inadvertent access and the undesirable behavior of connected systems.

Secure Development Environment

A secure development environment must be established to ensure that unneeded, unwanted and undocumented code is not introduced into the digital safety system – either operating system software or application software. Regulatory Positions 2.2 – 2.5 of of Revision 3 to RG 1.152 specifically identify controls that an applicant should implement during the development activities for safety related digital systems.

Platform Development

To initiate this evaluation, the NRC staff reviewed the Common Q Platform Topical Report (TR) WCAP-16097-P-A (ADAMS Accession No. ML031830959; public version ADAMS Accession No. ML031820484), and associated NRC staff safety evaluation. The Common Q platform underwent commercial grade dedication for use in safety-related applications. At the time of the original dedication, the QNX operating system on the Flat Panel Display System and the software resident in the AC160 controller were included in the scope of the dedication effort. Section 4.2 of the NRC staff's safety evaluation found the commercial grade dedication actions acceptable per EPRI-106439 for both the Flat Panel Display System and AC160 software.

Section 11.3 of the Common Q Platform TR indicates that the commercially dedicated platform is subject to configuration management control. The software configuration management process was initially described as part of the Common Q Platform TR with further detail contained in Section 6 of the Software Program Manual for Common Q Systems (WCAP-16096-NP-A, ADAMS Accession No. ML050350234). Section 4.3.1.k of the NRC staff safety evaluation for the Common Q Platform TR found that the Common Q Software Configuration Management Plan conformed to the provisions of IEEE Std. 828-1990, "IEEE Standard for Software Configuration Management Plans," and IEEE Std. 1042-1987, "IEEE Guide to Software Configuration Management," as endorsed by RG 1.169, "Configuration Management Plans for Digital Computer Software Used in Safety Systems of Nuclear Power Plants," issued September 1997.

Per WCAP-17427-P (ADAMS Accession No. ML11257A061; public version ADAMS Accession No. ML11257A050), the software and tools associated with the AC160 component are under the ASEA Brown Boveri (ABB) configuration management program. All changes are jointly approved by ABB and Westinghouse Electric Corporation (Westinghouse). Field-reported errors for the AC160 component are handled under the Westinghouse corrective action process and are assessed quarterly for impacts on applications that are fielded or under development.

In response to the NRC staff's RAIs on the PAMS application for WBN Unit 2, TVA submitted WCAP-17266-P, "Common Q Platform Generic Change Process" (ADAMS Accession No. ML102290177; public version ADAMS Accession No. ML102290176). The document describes the Common Q platform vendor's "10 CFR 50.59-like" process for evaluating all potential hardware and software changes to the Common Q platform. This document was developed in 2010, after the dedication of the Common Q platform, thus it was not reviewed by the NRC staff as part of the initial platform review. Since this review is focused on the WBN Unit 2 PAMS, WCAP-17266-P was not evaluated for generic applicability to the Common Q platform. However, the NRC staff did review the document as part of this evaluation and concluded that WCAP-17266-P does contain controls that, in addition to the previously reviewed software configuration control provisions, would further enhance the vendor's ability to preclude the introduction of unneeded or unwanted functionality. Software changes evaluated include those that would affect either the QNX or AC160 software.

The Common Q vendor collects operating experience on the fielded platforms under its corrective action process (WCAP-17427). On an annual basis, reports are evaluated to determine whether there are any errors that could impact any current applications. Changes made to the software for the Common Q platform since its dedication are captured in Section 2.2.2 of WNA-LI-00058-WBT-P, "Tennessee Valley Authority Watts Bar Unit 2 Post-Accident Monitoring System Licensing Topical Report," Revision 3 (ADAMS Accession No. ML110950334; public version ADAMS Accession No. ML110950333). Based upon the vendor configuration controls and evaluation and tracking of changes made to the dedicated platform, the NRC staff concludes that there is reasonable assurance that the developer of the platform software has taken appropriate steps to ensure platform reliability by guarding against introduction of unwanted, unneeded or undocumented code.

The QNX release for the platform has a cyclic redundancy check (CRC) stamp associated with the specific version of the operating system that is to be installed. In response to a RAI regarding use of the CRC stamp for verification of the QNX installation for the WBN Unit 2 PAMS, TVA provided a letter dated March 2, 2012 (ADAMS Accession No. ML12073A224), which cites use of a specific PAMS software installation procedure, WNA-IP-00528-WBT, that includes a CRC validation of the installed QNX software. In addition, the licensee notes that the execution of the installation procedure was witnessed by their independent V&V personnel.

The NRC staff asked a similar question regarding the installation of the AC160 software. In its March 2, 2012, response (ADAMS Accession No. ML12073A224), TVA indicated that the AC160 software installation validation was handled in a manner consistent with the CRC stamp approach described for the QNX software.

Based on the tools used to identify the platform software and the procedures applied to verify the identity of the software installed, the NRC staff concludes that there is reasonable assurance that the developer of the PAMS software has taken appropriate steps to ensure high functional reliability of the platform by administering controls to ensure that the correct platform software versions were installed in WBN Unit 2 PAMS.

In addition to the above measures, the NRC staff inquired whether the "generic" Common Q platform contains any installed features that are not strictly necessary for the WBN Unit 2 PAMS application. In its March 2, 2012, RAI response (ADAMS Accession No. ML12073A224), TVA clarified that the AC160 uses an element library from which logic elements are selected to implement that PAMS. An analysis was performed of the unused AC160 elements. In TVA's RAI response, it was concluded in WNA-AR-00051-GEN, Revision 0, "AC160 Unused PC Elements Non-Interaction Analysis," that unused elements of the AC160 are not reachable by the code and will not interact with the application.

Based on the original dedication of the Common Q platform, the software processes and procedures in place to perform software configuration management, new processes to further evaluate changes to the platform software, the tracking of all changes made since dedication, the controls in place to validate the correct version of platform software was installed, and analyses performed to ensure that superfluous code is not able to impact safety function, the NRC staff concludes that there is reasonable assurance that the developer of the platform software for the WBN Unit PAMS has taken appropriate steps to ensure high functional reliability of the platform by administering controls to protect the PAMS from the introduction of superfluous code or functions since its initial commercial grade dedication.

Application Development

For the Common Q PAMS, Revision 1 to WCAP-17427-P describes controls applied to the system development. The NRC staff reviewed this document and noted its consistency with APP-GW-J0R-012, "AP1000 Protection and Safety Monitoring System (PMS) Computer Security Plan," Revision 1, which was submitted in support of the AP1000 design certification review. [Note that the NRC staff evaluation of APP-GW-J0R-012 appears in Section 7.9.5 of "NUREG-1793, Supplement 2: Final Safety Evaluation Report Related to Certification of the AP1000 Standard Plant Design Docket No. 52-006" (ADAMS Accession No. ML112061231). The evaluation found APP-GW-J0R-012 to be acceptable.]

The Common Q PAMS vendor document specified that, during design and development, controls were focused on maintaining confidentiality (i.e., preventing loss of security-relevant information regarding the system) and integrity (i.e., preventing inadvert change to the system). The NRC staff finds that these goals are acceptable for establishing a secure development environment.

In TVA's March 2, 2012, RAI response, Westinghouse confirmed that WCAP-16096-NP-A, as augmented by WNA-LI-00058-WBT-P, was used to establish controls on the PAMS application development environment and processes. This approach is consistent with commitments made in Section 2.2.3.1.1.b of WCAP-17427-P.

In TVA's March 2, 2012, RAI response, Westinghouse clarified what had been stated in WCAP-17427-P, that all design, independent verification and validation (IV&V), and testing documentation associated with the Common Q PAMS development for WBN Unit 2 is maintained under the Westinghouse Quality Management System (QMS). Per Section 2.2.3.1 of WCAP-17427-P, QMS documents are stored in password-protected repositories (i.e., the Enterprise Document Management System (EDMS)), and all changes to controlled documents are performed per process requiring revision controls and multiple reviews and signatures.

The PAMS vendor also confirmed (as part of the March 2, 2012, RAI response) that WCAP-17427-P commitments regarding code review of both the Flat Panel Display application software and AC160 software were performed by the IV&V organization. References to these reviews were provided and their inclusion in the aforementioned EDMS was noted.

Section 2.3.1.5 of WCAP-17427-P addresses verification of security requirements. TVA's March 2, 2012, RAI responses also confirmed that the security-specific requirements for the PAMS application were verified as part of the overall system V&V effort. Revision 7 to WNA-VR-00283-WBT (ADAMS Accession No. ML12073A394; public version ADAMS Accession No. ML12073A392), contains the overall IV&V report. In its March 2, 2012, RAI response, TVA also referenced other software release records. In addition, the RAI response referenced the requirements traceability matrix for the PAMS (Revision 5 to WNA-VR-00279-WBT). The NRC staff concludes that tracing the requirements through design, implementation and test, combined with implementation of an IV&V program, consistent with NRC guidance, substantially supports the conclusion that only desired features and code were included in the final application.

The Software Program Manual (SPM) for the Common Q system (ADAMS Accession No. ML050350234) addresses configuration control for Common Q product developments. Any changes to software must go through a formal change process, which includes documentation and approval of changes, as well as regression analyses. The SPM also describes requirements traceability and analysis, independent code and document reviews and validation testing. In its March 2, 2012, RAI response (ADAMS Accession No. ML12073A224), TVA confirmed that these activities were performed.

Based on the development of the PAMS application in a controlled environment using appropriate processes for safety-related systems, which included traceability of requirements through design, implementation and test, as well as implementation of protections on code and relevant documentation, the NRC staff concludes that there is reasonable assurance that the PAMS application developer has taken appropriate measures to ensure high functional reliability by protecting the application from inclusion of unwanted, unneeded, and undocumented code and functions.

Secure Operational Environment

A secure operational environment must be established to ensure that predictable, non-malicious events will not degrade the reliable performance of the safety system. Regulatory Positions 2.1 through 2.5 of Revision 3 to RG 1.152 specifically identify analyses and associated design activities that should be addressed during system development. To support this evaluation, the NRC staff asked TVA a number of questions to clarify the rationale for and details regarding design attributes noted in the licensee's submittals.

The licensee stated that access to the PAMS requires access to either PAMS panels in the auxiliary instrument room or the main control boards where the maintenance and test panel (MTP) and an operators module (OM), respectively, are provided to allow operators to interact with the PAMS. The PAMS has a digital connection to the plant ICS.

Protection from Inadvertent Access

The licensee has stated that the PAMS has two human interface terminals that may read or send information to and from the PAMS components. The OM is located in the WBN Unit 2 main control room. The MTP is located in a locked cabinet in the auxiliary instrument room. WBN Unit 2's SDOE document notes that both of these locations are within the vital area where personnel access is restricted (ADAMS Accession No. ML11257A050, Attachment 9). The licensee and vendor identified these two points of access as requiring controls and design considerations to preclude inadvertent access to the components of PAMS performing the safety function.

The licensee stated that it will implement controls for making software changes, as needed. Software changes can be made to the PAMS from the MTP. WCAP-17427-P identifies the following access controls:
- The MTP cabinet requires an access key to unlock.
  - Note that Technical Instruction (TI), TI-12.09, "Plant Key Control," and Periodic Instruction, 0-PI-OPS-12.0, "Key Accountability Verification," (ADAMS Accession No. ML11257A050, Attachment 9), govern the control of keys at WBN2.
- An operator notification triggers when the PAMS cabinet door is open.
- A function enable keyswitch is required for any calibration or system changes.
- An operator notification triggers when the function enable keyswitch is placed in the "enable" position.
- A software load key is required for any software updates to the AC160 component. When the key is enabled, the MTP software will become non-operational. Note that the

function enable and software load enable keyswitches are keyed differently (ADAMS Accession No. ML11257A050, Attachment 9).

- An operator notification will trigger when the software load key is being used on the MTP to load software onto the AC160.
- Privileged user access to the QNX software is password protected.
- CRC checks are implemented on the FPDS and AC160 components for file systems and memory. Any detected changes during operations will produce alerts on the MTP and OM and will be sent to the ICS.
- AC160 processor load is monitored and will alert the operator upon exceeding a high limit for load.

Given the licensee's docketed materials describing the multiple layers of access control and operator notifications provided when any barrier to system changes is accessed, the NRC staff concluded that there is reasonable assurance that the design of the PAMS MTP promotes high functional reliability because the MTP will be protected from inadvertent access.

The function enable keyswitch that can be installed on the OM is not permanently installed. Thus, changes cannot be made from the OM during operations. Any maintenance performed on the OM would entail additional equipment and would occur outside of normal operations (ADAMS Accession No. ML12018A213). In addition, the OM is located in the main control room such that anyone accessing the OM during operations would be noticed. The NRC staff concludes that the OM is reasonably protected from inadvertent access events that could impact the high functional reliability of the PAMS.

Protection from Undesirable Behavior from Connected Systems

As stated in the licensee's submittals and clarified in TVA's December 22, 2011, RAI response (ADAMS Accession No. ML12018A213), consideration was given to minimizing (digital) connections to the WBN Unit 2 PAMS. As currently implemented, the only digital connection to the PAMS is between the PAMS MTP PC Node Box and the ICS. A connection between the WBN Unit 1 PAMS OM PC Node Box and the ICS was omitted in the WBN Unit 2 design.

Two isolation devices are used to prevent information from being transmitted from the ICS to the PAMS MTP PC Node Box. The first barrier is a hardware data diode described in the licensee's submittal. The second barrier consists of software functions on the MTP PC Node Box designed to limit communication into the MTP PC Node Box. WCAP-17427-P notes that the MTP itself does not perform a safety function for the PAMS (ADAMS Accession No. ML11257A061; public version ADAMS Accession No. ML11257A050). The MTP only accepts anticipated signals necessary to send communications to the hardware data diode. NRC staff reviewed this configuration with respect to Interim Staff Guidance (ISG)-04. The NRC staff evaluation is found in Section 7.5.2.2.3.7.1 of SSER 23 (ADAMS Accession No. ML11206A499). The NRC staff determined that the WBN Unit 2 PAMS connection to the ICS was consistent with the provisions of ISG-04.

A data storm test was performed on the MTP interface with the ICS to test the ability of these devices to prevent excessive traffic from the ICS from impacting PAMS functions. WNA-TR-02426-WBT, Revision 1, "Post-Accident Monitoring System Data Storm Test Report," documents the results of the testing. The testing confirmed that the MTP performed its isolation function without impacting PAMS safety-related functions (ADAMS Accession No. ML11257A050, Attachment 9).

Based upon the minimization of digital connections to other systems and the isolation measures taken and confirmed by test for the remaining digital connection, the NRC staff concludes that there is reasonable assurance that the WBN Unit 2 PAMS will retain its high functional reliability in service because it is protected from the undesirable behavior of connected digital systems.

CONCLUSION

Based upon its review of the information provided by TVA, the NRC staff concludes that a Secure Development and Operational Environment has been established for the WBN Unit 2 Common Q PAMS which is consistent with the regulatory positions found in RG 1.152, Revision 3. This conclusion is based upon the controls placed on the pre-developed Common Q platform, the controls on the development of the PAMS application and the WBN Unit 2 facility and PAMS design features, including the controls to preclude two-way communication to other systems. Therefore, the NRC staff concludes that the PAMS has been designed and will be operated with provisions for ensuring its high functional reliability and therefore meets the requirements 10 CFR 50.55a(h), GDC 21 of Appendix A to 10 CFR Part 50, and Criterion II of Appendix B to 10 CFR Part 50.

This safety evaluation **closes Open Item 98**, as identified in Section 7.5.2.2.3.12 of SSER 23 (ADAMS Accession No. ML11206A499).

7.5.2.3.4   Technical Evaluation

Open Item 77

Open Item 77 states, "It is unclear to the NRC staff which software V&V documents are applicable to the high radiation containment area radiation (HRCAR) monitors. TVA should clarify which software V&V documents are applicable, in order for the staff to complete its evaluation."

By letters dated September 1, 2011 (ADAMS Accession No. ML11257A048), and November 14, 2011 (ADAMS Accession No. ML11322A099), TVA responded to Open Item 77. In these letters TVA clarified the document number and the version of each of the software V&V test documents (Versions V 1.0, V 1.1, and V 1.2) that are applicable to the WBN Unit 2 HRCAR monitors.

By letter dated October 13, 2011 letter (ADAMS Accession No. ML11291A095), TVA responded to a followup question regarding the quality control of the V&V procedures. The NRC staff wanted to confirm that the procedures used for V&V testing were prepared in accordance with Appendix B to 10 CFR Part 50. In its October 13, 2011, letter, TVA responded that the procedures were prepared in accordance with Criterion V, "Instructions, Procedures, and Drawings," of Appendix B to 10 CFR Part 50 and that they contained appropriate qualitative and quantitative acceptance criteria to assure that important activities have been satisfactorily accomplished. Based on the review of the information provided by this TVA letter and other RAI responses, the NRC staff concludes that the TVA response is satisfactory.

Based on the review of TVA's October 13, 2011, and other RAI responses, the NRC staff concludes that TVA provided sufficient information to confirm which versions of the software V&V documents apply to WBN Unit 2. Based on the satisfactory review of the TVA responses, the NRC staff has **closed Open Item 77**.

Open Item 81

Open Item 81 states, "The extent to which TVA's supplier, General Atomics (GA), complies with EPRI TR-106439 and the methods that GA used for its commercial dedication process should be provided by TVA to the NRC staff for review."

By letter dated September 30, 2011 (ADAMS Accession No. ML11287A254), TVA provided the following response to this open item:

"Compliance with EPRI TR-106439 and the methods GA uses for its commercial dedication process are documented in GA procedure OP-7.3-240. Attachment 1 contains proprietary GA procedure OP-7.3-240, Revision K, Safety-Related Commercial Grade Item Parts Acceptance."

This procedure (OP-7.3-240) is based on EPRI-NP5652 "Guidelines for the Utilization of Commercial Grade Items in Safety Related Applications," EPRI-TR102260, "Supplemental Guidance for the Application of EPRI Report NP-5652 on the Utilization of Commercial Grade Items", and EPRI TR-106439, "Guideline on Evaluation and Acceptance of Commercial Grade Digital Equipment for Nuclear Safety Applications." NRC staff review determined that the procedure (OP-7.3-240) is acceptable.

In the October 13, 2011, letter (ADAMS Accession No. ML11291A095), TVA submitted a "White Paper" that included a sample package that demonstrated how the commercial grade dedication was accepted by GA for an AC filter unit (Isotrol+105 filter) that is a subcomponent in the RM-1000 radiation monitors. The filter unit is a commercial grade subcomponent that is dedicated in accordance with GA procedure OP-7.3-240. This package included a receipt inspection, a certificate of conformance, a bill of lading completed by the item supplier to confirm the shipped item model number and the customer purchase order number, a database change request for replacing the component in the assembly unit, critical characteristics acceptance plan, commercial grade item engineering evaluation sheets to verify the critical characteristics, and item supplier data sheets. GA also provided a copy of the supplier's full scope audit report to confirm that the supplier meets the requirements of International Standardization Organization 9001, as specified by the assigned quality level approved by GA. (It should be noted that GA has used the abbreviations GA as well as GA-ESI where ESI stands for Electronics Systems Inc., which is an affiliated company of GA.)

After reviewing the documents by TVA, the NRC staff observed that the package did not include the functional test results for the AC filter. By letter dated November 18, 2011 (ADAMS Accession No. ML113130218), the NRC staff requested TVA to provide the documentation for the functional test results to enable the NRC staff to complete its evaluation of this package. If functional test results were not available, then TVA was to provide the justification for the lack of availability and to provide the complete commercial dedication package for an alternate component for NRC Staff review.

In a letter dated December 22, 2011 (ADAMS Accession No. ML12018A213), TVA informed NRC that the functional tests were performed using an automated test machine that only gives the result as to whether the component passed or failed the test and provided an alternate commercial grade dedication package for a 24 VDC, 1.8 Amp power supply. This package contained a GA receipt inspection, a bill of lading that is also a certificate of compliance to confirm that the item supplied is correct per the purchase order, a data base change request for incorporating the item, critical characteristics acceptance plan, a commercial grade item engineering evaluation performed by GA, a quality control inspection report by GA that includes the scope of supply, a visual inspection, the test equipment description, the test setup, the test procedure listing various tests and acceptance/rejection criteria. Further test reports of all of the 24 power supplies were also included in this package, which confirmed that the functional test results were satisfactory.

In Enclosure 2 to its letter dated (ADAMS Accession No. ML12116A151), TVA confirmed that the test machine used for testing the AC filter unit (Isotrol+105 filter) produces only a pass/fail test. Enclosure 2 to the April 23, 2012, letter, includes the test setup and confirms that the test was a pass/fail test. The NRC staff reviewed the test setup and confirmed that the test was indeed a pass/fail test.

With the submittals noted above, TVA resolved the issue of the AC filter and provided adequate documentation to satisfy the commercial grade dedication requirements for the power supply per the commercial dedication procedure OP-7.3-240, Rev. K, which is based on the guidance of EPRI NP-5652, EPRI TR-102260, and EPRI TR-106439.

TVA has provided sufficient information with its response to confirm that the test machine for the AC filter is indeed a pass/fail type of test device. Furthermore, TVA provided a complete commercial grade dedication package for an alternate component (a power supply) to confirm that the commercial grade dedication was carried out in accordance with the GA procedure. Based on the TVA response, NRC staff considers **Open Item 81 closed**.

7.7   Control Systems Not Required for Safety

7.7.1   System Description

7.7.1.4   Distributed Control System

The NRC staff's evaluation of the Foxboro Intelligent Automation (I/A) distributed control system (DCS) was documented in SSER 23, Section 7.7.1.4. The NRC staff asked TVA to confirm whether the Foxboro I/A system was developed under a 10 CFR Part 50, Appendix B compliant program. While some parameters listed in RG 1.97, "Criteria for Accident Monitoring Instrumentation for Nuclear Power Plants," Revision 4, issued June 2006, pass through the Foxboro I/A system, not all RG 1.97 parameters are required to be safety-related. TVA responded in Item No. 8 of its letter dated April 15, 2011 (ADAMS Accession No. ML11136A053). TVA stated that, "Foxboro I/A is a non-safety-related system. Therefore, 10 CFR [Part] 50 Appendix B is not applicable." The NRC staff confirmed that TVA's response is acceptable, because the RG 1.97 variables processed by the Foxboro I/A system are not safety-related, as shown in WBN Unit 2 FSAR Table 7.5-2, "Regulatory Guide 1.97 Post Accident Monitoring Variables Lists Legend."

7.7.1.9   In-Core Instrumentation System

Disposition of Open Items (Appendix HH)

Open Item 118

Open Item 118 states, "TVA should provide to the NRC staff a description of how the other vanadium detectors within the IITA [In-core Instrumentation Thimble Assembly] would be operable following the failure of an SPND [Self-Powered Neutron Detector]."

By letter dated September 30, 2011 (ADAMS Accession No. ML11287A254, Item No. 2 of Enclosure 1), TVA provided a response to this action item. In its response, TVA explained that each SPND element contains its own Mineral Insulated (MI) cable, which physically and electrically isolates each detector element from all the other elements inside the IITA. Therefore, there is no direct link between the measured signals from individual detectors inside an IITA. Consequently, there is no reason for the failure of one detector element to affect the operability of any of the other elements within the same IITA.

In its letter dated October 14, 2011 (ADAMS Accession No. ML11291A095, Item No. 3 of Enclosure 1, and Attachment 6), TVA forwarded a copy of a letter from Westinghouse Electric Corporation providing additional detail regarding how the BEACON software uses the data from SPNDs in an IITA in which one or more SPND elements have failed. The Westinghouse letter explained that the approved Westinghouse WCAP-12472, "BEACON Core Monitoring and Operation Support System," Addendums 1 (ADAMS Accession No. ML003678190) and 2 (ADAMS Accession No. ML021270086) describe the uncertainty and methodology used to establish the number and distribution of required SPND sensors and how the system identifies data associated with a failed SPND. In particular, Westinghouse explained that the WINCISE system is considered operable provided that the total number of operable SPNDs is greater than a particular parameter specified in the model. For example, the WINCISE system is considered to be operable when the outputs of the SPNDs read are above certain percentage signal value and the number of operable SPNDs in the top and bottom halves of each quadrant exceeds a certain limit, and the total number of operable SPNDs in every quadrant meets the criteria established for that quadrant. Furthermore, the IITA design features five SPNDs of sequentially increasing length that overlap the measurement range. Thus, the BEACON system is able to adequately function using information from all available sensors, but with different uncertainty depending on which sensors have failed. Based on these responses, the NRC staff finds that this action item has been adequately addressed, and considers **Open Item 118 closed**.

Open Item 120

Open Item 120 states, "TVA should confirm to the NRC staff that the maximum over-voltage or surge voltage that could affect the system is 264 VAC, assuming that the power supply cable to the SPS cabinet is not routed with other cables greater than 264 VAC."

By letter dated September 1, 2011, (ADAMS Accession No. ML11257A048, Item No. 7 of Enclosure 1) TVA responded to this action item. By letter dated October 14, 2011 (ADAMS Accession No. ML11291A095), TVA provided additional information and confirmed that the power supply cable to the WINCISE Signal Processing System (SPS) cabinet is not routed with any cables with greater than the maximum steady-state voltage allowed in Westinghouse technical note WNA-CN-00157-WBT-P, Revision 0, "Watts Bar 2 IIS [in-core instrumentation system] Signal Processing System Isolation Requirements." In this letter, TVA committed to "lock" cable routing in the Integrated Cable and Raceway Design System (ICRDS) to prevent cables greater than the maximum steady voltage allowed in WNA-CN-00157-WBT-P from being routed in the future with the SPS cabinet power supply cables. TVA confirmed in its December 22, 2011 letter (ADAMS Accession No. ML12018A213, Item No. 7), that the cable routing has been locked within the ICRDS. Further, during an audit to review Westinghouse documents, the NRC staff reviewed Westinghouse letter WBT-TVA-1060. This letter describes the power supply requirements for the WINCISE SPS cabinet and includes an external power supply diagram, which showed the planned cable routing. Based on this information, the NRC staff finds that this action item has been adequately addressed, and considers **Open Item 120 closed**.

Open Item 121

Open Item 121 states, "TVA should submit the results to the NRC staff of a 600 VDC dielectric strength test performed on the IITA assembly."

Westinghouse report WNA-CN-00157-WBT, "Watts Bar 2 Incore Instrumentation System Signal Processing System Isolation Requirements," Revision 0 (ADAMS Accession No. ML11257A048), requires the IITA assembly and MI connecting cabling from the IITA to the point where the core exit thermalcouple (CET) cabling is separated to be at a certain voltage level. This MI portion is identified as the WINCISE "1 to 2 Transition Cable Assemblies."

In a letter dated October 14, 2011 (ADAMS Accession No. ML11291A095), TVA submitted a copy of Westinghouse Report WBT-D-3518 P, "Partial Response to Item 9 SSER Item 121 Testing of Completed WINCISE 1 to 2 Transition MI Cable Assemblies - Watts Bar 2," dated October 2011 (Item No. 5 of Enclosure 1, and Attachment 9). This report summarized the results of the dielectric testing performed to all WINCISE 1 to 2 Transition Cable Assemblies, which demonstrated that the WINCISE 1 to 2 Transition MI Cable Assemblies passed the identified testing program. Furthermore, to evaluate the WINCISE system, during an audit, the NRC staff reviewed the letter from R. W. Morris to D. Menard, LTR-ME-10-3, "Watts Bar 2 Incore Instrumentation System Dielectric Characteristics of Completed MI Cable Assemblies," dated January 11, 2010. This letter summarized the evaluation performed for the MI cable to be capable of withstanding a specific over-voltage and surge voltage level. Based on this information, the NRC staff concluded that the information submitted to document the testing and evaluation for the MI cable assembly is acceptable. The NRC staff requested that TVA provide the results of the testing performed on the IITA for NRC review and verification.

By letters dated November 14 and 30, 2011 (ADAMS Accession No. ML11322A099 and ML11341A156, respectively), TVA forwarded a copy of a Westinghouse letter explaining that the cable manufacturer performed type-testing of the IITA. The result from this testing is reported in IST Report #021-8557. This result was reflected in Revision 1 of Westinghouse's technical note WNA-CN-00157-WBT, "Watts Bar 2 Incore Instrumentation System Signal Processing System Isolation Requirements," submitted to the NRC as Attachment 6 of TVA's letter dated November 30, 2011. In addition, during an audit to review Westinghouse documents, the NRC staff reviewed a copy of the Westinghouse test report and confirmed that the IITA assembly did not exhibit unacceptable resistance breakdown during testing and passed the resistance test.

In Attachment 2 of TVA's letter dated November 14, 2011, TVA explained Westinghouse's approach to estimate the maximum IITA cable leakage resistance at the minimum temperature for criticality. Further, in TVA's letter dated November 30, 2011 (ADAMS Accession No. ML11341A156), TVA submitted a copy of Revision 1 of Westinghouse's WNA-CN-00157-WBT, which included a summary of the IITA cable leakage resistance. In addition, during an audit to review Westinghouse documents, the NRC staff reviewed Westinghouse and TVA documents that confirmed the resistance for the IITA cable at the minimum temperature for criticality. Based on this information, the NRC staff found that the core exit thermocouples have a dielectric voltage rating of at least 600 V, and that the estimated cable leakage resistance met the acceptance criteria established in the Design Specification for the IITA cable. Therefore, TVA and Westinghouse have demonstrated the ability of the IITA assembly to maintain insulation dielectric integrity and withstand potential brief overvoltage. This supports the NRC staff's ability to make a determination that the proposed design meets the requirements for isolation discussed in RG 1.75, "Physical Independence of Electric Systems," which endorses IEEE 384-1981, "IEEE Standard Criteria for Independence of Class 1 E Equipment and Circuits" for ensuring sufficient independence between safety and non-safety components. Based on the information provided by TVA, the NRC staff finds that this action item has been adequately addressed, and considers **Open Item 121 closed**.

Open Item 123

Open Item 123 states, "TVA should provide an explanation to the NRC staff of how the system will assign a data quality value to notify the power distribution calculation software to disregard data from a failed SPND."

By letter dated September 30, 2011 (ADAMS Accession No. ML11287A254, Item No. 3 of Enclosure 1), TVA responded to this action item by explaining how the WINCISE system performs data validation. In particular, TVA explained that the BEACON Data Processing (BDP) application software continuously reviews detector signal measurements; performs on-demand checks of signal cable leakage resistance; and performs validity checks of this information relative to the defined maximum change in the measured signal between two measurement intervals, high and low current limits, and minimum acceptable cable leakage resistance measurement limits. If any of these criteria are violated, the BDP application sets a data quality bit contained in the digital representation of the current value to "BAD." The BEACON System Plant Interface function automatically disregards the "BAD" data. Also, as described above in Open Item 118, the BEACON system would use information from only "available" sensors. Based on this response, the NRC staff finds that this action item has been adequately addressed, and considers **Open Item 123 closed**.

Open Item 125

Open Item 125 states, "TVA should provide clarification to the NRC staff of the type of connector used with the MI cable in Unit 2, and which EQ [environmental qualification] test is applicable."

By letter dated September 30, 2011 (ADAMS Accession No. ML11287A254, Item No. 4 of Enclosure 1), TVA responded to this action item by explaining that the electrical connectors used with the WBN Unit 2 WINCISE MI Cable Assemblies are glass-to-metal seal technology connectors fabricated for Westinghouse by Meggitt Safety Systems.

Furthermore, TVA noted that Westinghouse summarized the environmental qualifications for the MI cable and its connectors in Westinghouse Proprietary Report DAR-ME-09-10, "Qualification Summary Report for the WINCISE Cable and Connector Upgrade at Watts Bar Unit 2," which TVA submitted to the NRC by letter dated May 6, 2011 (ADAMS Accession No. ML11129A205). Also, in TVA's November 14, 2011, letter (ADAMS Accession No. ML11322A099), TVA explained the manufacturer performed a voltage breakdown test per MIL-STD-202, Method 301, "Dielectric Withstand Voltage." In TVA's letter dated November 30, 2011 (ADAMS Accession No. ML11341A156), TVA submitted a copy of Revision 1 of Westinghouse's WNA-CN-00157-WBT, which includes a reference to the test summary of the voltage breakdown test. Furthermore, during a teleconference held in December with the NRC staff, which is summarized in TVA's letter dated February 1, 2012 (ADAMS Accession No. ML12034A164), TVA and Westinghouse clarified that the 600 V dielectric testing was performed for the entire

IITA assembly, and that this test was performed successfully. In addition, during an audit to review Westinghouse documents, the NRC staff reviewed a Westinghouse letter that confirmed that all 58 WINCISE 1-to-2 cable were subjected and passed successfully the dielectric strength test (conductor to conductor and conductor to sheath basis). Based on this response, the NRC staff finds that this action item has been adequately addressed, and considers **Open Item 125 closed**.

Open Item 126

Open Item 126 states, "To enable the NRC staff to evaluate and review the IITA environmental qualification, TVA should provide the summary report of the environmental qualification for the IITA."

By letter dated September 30, 2011 (ADAMS Accession No. ML11287A254, Item No. 5 of Enclosure 1, and Attachment 2), TVA responded to this action item by summarizing the Westinghouse qualification for the Ex-Vessel Portion of the WINCISE IITA for WBN Unit 2. This report summarizes the environmental and seismic qualification for the WBN Unit 2 ex-vessel portion of the IITA, and explains how the IITA meets the requirements in IEEE 323 and IEEE 344.

The IITA contains two portions, an in-vessel and ex-vessel portion. The in-vessel portion of the IITA consists of self-powered vanadium detectors and thermocouple and is considered from the Swagelok nut to the end of the IITA (bullet nose). The ex-vessel portion of the IITA is considered to be the hermetically sealed portion of the assembly which extends from the Swagelok assembly to the weldable Electronic Resources Division (ERD) multipin electrical connector.

In the qualification summary report, Westinghouse explained that they qualified the IITA using similarity analysis. For each qualification testing, Westinghouse described how the ERD multipin electric connector met the requirement. Furthermore, Westinghouse explained that the tested ex-vessel portion of the IITA in Westinghouse (ABB CE [Combustion Engineering]) Report Number CE-NPSD-240-P, "Summary Report: Class 1E Qualification Test of the Incore Instrument (Core Exit Thermocouple Portion) and Mineral Insulated Cable Assembly," and ERD multipin electrical connector in Westinghouse (ABB CE) Report Number CE-NPSD-275-P, "Summary Report: Class 1E Qualification Test of the Electronic Resources Division (ERD) Electrical Connectors and Mineral Insulated Cable," both fulfill the electrical operability acceptance criteria throughout all phases of testing. The NRC staff reviewed these two Westinghouse documents during an audit.

The CE-NPSD-275-P summary report showed that the qualification testing meets or exceeds the Watts Bar Unit 2 requirements. Therfore, the WBN Unit 2 ex-vessel portion of the IITA with ERD multipin electrical connector is qualified for Class 1E application in accordance with the methodology and guidance of IEEE 323 and IEEE 344. Based on this information, the NRC staff finds that the qualification summary reports adequately address the qualification requirements for the ex-vessel portion of the IITA.

In its letter dated November 30 (ADAMS Accession No. ML11341A156), TVA provided an overview of the analysis of the environmental and seismic/structural qualification methodology and results contained in the report CE-NPSD-240-P, "Summary Report: Class 1E Qualification of the Incore Instrument (Core Exit Thermocouple Portion) and Mineral Insulated Cable Assembly" seismic qualification testing performed on the core exit thermocouple portion of the assembly. In addition, during an NRC staff audit to review Westinghouse documents, the NRC staff reviewed CE-NPSD-240. This document summarizes how the incore instrument (core exit thermocouple portion) and MI cable assembly meets their Class 1E qualification requirements. The IITA was tested as a complete assembly during the EQ testing. In TVA's letter March 2, 2012 (ADAMS Accession No. ML12065A142), TVA confirmed that the ex-vessel portion is representative of the IITA that will be installed at WBN Unit 2. The in-vessel portion used the same CET to be installed at WBN Unit 2. Because the CET is the only portion of the IITA that is safety related, the results reviewed by the NRC staff showed that the IITA met the specified acceptance criteria. Based on this, the NRC staff finds that this action item has been adequately addressed, and considers **Open Item 126 closed**.

Open Item 127

Open Item 127 states, "TVA should provide a summary to the NRC staff of the electro-magnetic interference/radio-frequency interference (EMI/RFI) testing for the MI cable electro-magnetic compatibility (EMC) qualification test results."

By letters dated September 30, 2011, November 14, 2011, and December 22, 2011 (ADAMS Accession Nos. ML11287A254, ML11322A099, and ML12018A213, respectively), TVA responded to this action item (Item No. 6 of Enclosure to the September 30, 2011, and November 14, 2011, letters, and Item No. 5 of Enclosure 1 to the December 22, 2011, letter) by explaining that the MI cable assemblies to be supplied for the WBN Unit 2 IIS were not subjected to a product-specific EMI/EMC test program, but rather were qualified by analysis of the design. Within the IITA, the CET is insulated with crushed Alumina (synthetically produced aluminum oxide, Al_2O_3) contained in an overall stainless steel tubular sheath. Each individual SPND consists of a Vanadium emitter wire, surrounded by crushed Alumina, which is surrounded by a grounded stainless steel tubular sheath. The thermocouple sheath, the SPND sheaths, and the overall IITA sheath are all electrically grounded at the reactor vessel providing a path to effectively divert EMI/RFI. The combination of the stainless steel sheath material joined to the stainless steel connectors provides for 100 percent shielding coverage.

The ex-vessel MI cable assembly consists of silicon dioxide (SiO_2) insulation, enclosing the SPND signal leads and core exit thermocouple lead wires, each one surrounded by a separate grounded stainless steel tubular sheath. The combination of the stainless steel sheath material joined to the stainless steel connectors provides for 100 percent shielding coverage. The exterior surfaces of the IIS MI Cable Assemblies are post-accident qualified, and as such, are required to be 100 percent hermetically sealed. This hermeticity of the MI Cable Assembly design and construction also demonstrates the absence of any apertures or seams that would compromise the shielding effectiveness of the assemblies, and thus providing the necessary protection against EMI/RFI interferences. To provide the necessary grounding of the MI cable, the cable assemblies are to be directly secured to seismically qualified in-containment cable supports at regular intervals along the length of the cable run. The frequency of this support arrangement provides multiple low impedance paths to ground for the cable assemblies to effectively divert EMI/RFI.

As an additional layer of protection, the IIS MI Cable Assemblies for WBN Unit 2 include an interlocked stainless steel material hose assembly that is brazed and seal welded to the electrical connector backshell at each end of the cable assemblies. This hose assembly provides supplemental EMC shielding protection for the signals.

In TVA's letters dated September 1, 2011 (ADAMS Accession No. ML11257A048, Attachment 2), and November 14, 2011 (ADAMS Accession No. ML11322A099), Westinghouse explained that the maximum current from a Vanadium detector is sufficiently low which, in the event of a short circuit from emitter to sheath within the cable, restricts the energy available to an amount that will preclude melting or other damage to the protective sheath. In case of breakage to the sheath, the detector leakage current will be shunted to common (plant ground) via the detector sheath. Further, the design maximum emitter current is sufficiently low that any short within the IITA will so restrict the energy available that further damage is precluded. Confirmation of electrical separation was important since the CET and SPDNs share the same IITA. Therefore, the NRC staff concludes that the dual barrier design combined with the low detector current provides inherent EMI/RFI protection for the MI cable, and that there is no credible fault originating within the non-1E portion of the MI cable can negatively impact the CET. This conclusion supports the NRC staff's ability to make a determination that the proposed design meets the requirements for isolation discussed in RG 1.75, which endorses IEEE 384-1981 for ensuring sufficient independence between safety and non-safety components. Based on this response, the NRC staff finds that this action item has been adequately addressed, and considers **Open Item 127 closed**.

Open Item 129

Open Item 129 states, "TVA should verify to the NRC staff resolution of the open item in WNA-CN-00157-WBT for the Quint power supply (to be installed in the SPS cabinet) to undergo EMC testing of 4 kV to validate the assumptions made in the Westinghouse analysis."

By letter dated September 30, 2011 (ADAMS Accession No. ML11287A254, Item No. 7 of Enclosure 1, and Attachment 10), TVA responded to this action item by explaining that the power supply to be installed in the SPS cabinet was tested as part of the WINCISE SPS. TVA submitted a copy of the Westinghouse Electric Company EQ-QR-39-WBT-NP, "Equipment Qualification Summary Report for WINCISE Signal Processing System," in Attachment 10. The test results provided in this summary report showed the WINCISE SPS cabinet, and thus the power supply, successfully complied with the emissions requirements of RG 1.180, "Guidelines for Evaluating Electromagnetic and Radio-Frequency Interference in Safety-Related Instrumentation and Control Systems," Revison 1, issued 2003. Based on this response, the NRC staff finds that this action item has been adequately addressed, and considers **Open Item 129 closed**.

9 AUXILIARY SYSTEMS

9.1 Fuel Storage Facility

Disposition of Open Items (Appendix HH)

9.1.3 Spent Fuel Pool Cooling and Cleanup System

Open Item 60

Open Item 60 states, "TVA should amend the FSAR description of the design and operation of the spent fuel pool cooling and cleanup system in FSAR Section 9.1.3 as proposed in its December 21, 2010, letter to the NRC."

The NRC staff confirmed that TVA updated the FSAR in Amendment 103 to add the description of the system as propsed in its letter dated December 21, 2010. Therefore, **Open Item 60 is closed**.

9.5 Other Auxiliary Systems

9.5.1 Fire Protection

During the operating licensing review for Watts Bar Nuclear Plant (WBN) Unit 1, the NRC staff documented its review of the WBN fire protection program in Appendix FF of NUREG-0847, "Safety Evaluation Report Related to the Operation of Watts Bar Nuclear Plant Units 1 and 2," Supplemental Safety Evaluation Report (SSER) 18, issued October 1995, and SSER 19, issued November 1995. As part of the operating license application for WBN Unit 2, TVA submitted the As-Designed Fire Protection Report (FPR) for WBN Units 1 and 2 to the NRC by letter dated December 18, 2010, as revised and supplemented by letters dated December 20, 2010; January 14, March 16 and 31, May 6, 18, and 26, June 7 and 17, July 1 and 22, August 5 and 15, September 30, October 28, November 21 and 30, 2011; March 13, April 12, 17, and 26, May 9 and 30, June 7 and 27, July 19, September 13, December 20, 2012; February 7 and 28, and March 13, 2013. The NRC staff's detailed evaluation of the updated fire protection program appears in Appendix FF to this SSER.

In the FPR, TVA stated that, "the purpose of the Fire Protection Report (FPR) is to consolidate a sufficiently detailed summary of the WBN regulatory required Fire Protection Program into a single document and to reflect the design as-constructed at the time of fuel load." The FPR describes the operational phase of the fire protection program. Accordingly, the NRC staff reviewed the entire fire protection program, except as noted below, using the agency's fire protection requirements and review guidance. Because WBN consists of two units of identical design, this evaluation applies to the fire protection program for both units, except as noted.

The NRC staff's review did not include Section 7, "Unit 1 Operator Manual Actions [OMAs]," of Part VII of the FPR. The NRC's approval of the WBN Unit 1 OMAs is documented in SSER 18.

In Staff Requirements Memorandum SECY-07-0096, "Possible Reactivation of Construction and Licensing Activities for the Watts Bar Nuclear Plant Unit 2," dated July 25, 2007, the Commission directed the NRC staff to use the existing WBN Unit 1 licensing basis as the reference basis for the WBN Unit 2 review. To that end, where applicable, the NRC staff used the WBN Unit 1 approvals, as documented in SSER 18 and 19, as the basis for its approvals in this evaluation instead of the agency's current guidance. The NRC used its current guidance as the basis for approval for WBN Unit 2 OMA, associated circuits, multiple spurious operation, fire water system design demand, the auxiliary control room, and radiant energy shields.

The NRC staff met with TVA on January 19, February 3 and 15, March 29, April 22, May 12, June 30, July 12 and 28, August 31, November 16, and December 21, 2011; and February 2, 2012, to discuss technical issues related to WBN's fire protection program and its implementation. The NRC staff also conducted an audit at WBN from October 25-27, 2011, which was documented by a report dated December 20, 2011 (ADAMS Accession No. ML113500239).

Unless otherwise noted, all information cited in the evaluation found in Appendix FF is from the WBN FPR dated March 13, 2013 (ADAMS Accession No. ML130840169).

On the basis of its review of TVA's as-designed FPR and TVA's supplemental information as referenced by this evaluation, the NRC staff concludes that the fire protection program for WBN, with the exception of Unit 1 specific OMAs, meets 10 CFR 50.48(a) and GDC 3 of Appendix A to 10 CFR Part 50, and is consistent with Sections III.G, III.J, III.L, and III.O of Appendix R to 10 CFR Part 50 and Appendix A to BTP (APCSB) 9.5-1, May 1976, with properly justified deviations and exceptions. Therefore, the NRC staff finds the as-designed FPR acceptable, contingent on the completion of the confirmatory items identified in Section 8.0 of this evaluation **(Open items 140, 141, 142, and 143, Appendix HH)**. NRC approval of the Unit 1 OMAs is documented in SSER 18, October 1995, of NUREG-0847, "Safety Evaluation Report Related to the Operation of Watts Bar Nuclear Plant Units 1 and 2."

12 RADIATION PROTECTION

Disposition of Open Items (Appendix HH)

12.4 Radiation Protection Design Features

Open Items 112, 113, and 114

Open Item 112 states, "TVA should provide an update to the FSAR reflecting the radiation protection design features descriptive information provided in its letter dated October 4, 2010."

Open Item 113 states, "TVA should provide an update to the FSAR reflecting the justification for the periodicity of the COT [channel operability test] frequency for WBN nonsafety-related area radiation monitors."

Open Item 114 states, "TVA should update the FSAR to reflect that WBN meets the radiation monitoring requirements of 10 CFR 50.68."

TVA provided the requested updates in FSAR Amendment 105, and the NRC staff verified that the updates were acceptable. Therefore, **Open Items 112, 113, and 114 are closed**.

12.6 Health Physics Program

Open Item 116

Open Item 116 states, "TVA should update the FSAR to reflect the qualification standards of the RPM [radiation protection manager] as provided in its letter to the NRC dated October 4, 2010." TVA provided the requested update in FSAR Amendment 105, and the NRC staff verified that the update was acceptable. Therefore, **Open Item 116 is closed**.

15   ACCIDENT ANALYSIS

15.2   Normal Operation and Anticipated Transients

15.2.2 Increased Cooling Transients

15.2.4.4   Chemical and Volume Control System Malfunction That Results in a Decrease in Boron Concentration in the Reactor Coolant

Section 15.2.4, "Uncontrolled Boron Dilution," WBN Unit 2 Final Safety Analysis Report (FSAR) contains TVA's analysis of a chemical and volume control system (CVCS) malfunction that results in a decrease in boron concentration in the reactor coolant.

Regulatory Evaluation

A CVCS malfunction that results in a decrease in boron concentration in the reactor coolant (also known as B dilution) is an anticipated operational occurrence (AOO) and is classified as an American Nuclear Society Condition II event.

Unborated water can be added to the reactor coolant system (RCS) through the CVCS. This may happen inadvertently because of operator error or CVCS malfunction, and cause an unwanted increase in reactivity and a decrease in shutdown margin. The operator should stop this unplanned dilution before the shutdown margin is eliminated.

In its review, the NRC staff used the guidance in NUREG-0800, "Standard Review Plan for the Review of Safety Analysis Reports for Nuclear Power Plants: LWR Edition" (SRP), Chapter 15, "Transient and Accident Analysis," Section 15.4.6, "Inadvertent Decrease in Boron Concentration in the Reactor Coolant System (PWR)," Revision 2, issued March 2007. The NRC staff's review covered (1) conditions at the time of the unplanned dilution, (2) causes, (3) initiating events, (4) the sequence of events, (5) the analytical model used for analyses, (6) the values of parameters used in the analytical model, and (7) results of the analyses.

The NRC based its acceptance criteria, in part, on the following regulatory requirements:

- General Design Criterion (GDC) 10, "Reactor Design," of Appendix A, "General Design Criteria for Nuclear Power Plants," to 10 CFR Part 50, requires, in part, licensees to design the reactor core and associated coolant, control, and protection systems with appropriate margin to ensure that specified acceptable fuel design limits (SAFDLs) are not exceeded during any condition of normal operation, including the effects of AOOs

- GDC 15, "Reactor Cooolant System Design," of Appendix A to 10 CFR Part 50 requires, in part, licensees to design the RCS and associated auxiliary, control, and protection systems with sufficient margin to ensure that the design conditions of the reactor coolant pressure boundary (RCPB) are not exceeded during any condition of normal operation, including AOOs

- GDC 26, "Reactivity Control System Redundancy and Capability," of Appendix A to 10 CFR Part 50 requires, in part, licensees to provide a reactivity control system that can reliably control reactivity changes to ensure that, under conditions of normal operation, including AOOs, SAFDLs are not exceeded.

<u>Technical Evaluation</u>

In its analysis of Condition II events in PWRs, including the B dilution event, the NRC staff uses, in part, the following three specific acceptance criteria, as described in SRP Section 15.4.6, in determining whether the applicant has satisfied the regulatory requirements:

(1) Pressure in the RCS and main steam system should be maintained below 110 percent of the design values in accordance with the American Society of Mechanical Engineers (ASME) Boiler and Pressure Vessel Code.

(2) Fuel cladding integrity shall be maintained by ensuring that the minimum departure from nucleate boiling ratio (DNBR) remains above the 95/95-DNBR safety limit (i.e., there will be at least a 95-percent probability that departure from nucleate boiling (DNB) will not occur on the limiting fuel rods during normal operation, operational transients, or any transient conditions arising from faults of moderate frequency (Condition I and II events) at a 95-percent confidence level).

(3) An AOO should not generate a postulated accident without other faults occurring independently or result in a consequential loss of function of the RCS or reactor containment barriers.

The B dilution analysis is performed principally to show that the second criterion is met. It is basically a time-dependent reactivity balance to determine whether enough time is available, for automatic or manual actions, to prevent the loss of all shutdown margin (i.e., to prevent dilution to criticality). If the core does not become critical, then second criterion will be satisfied (i.e., the minimum DNBR will remain above the 95/95 limit).

In its application, TVA did not explain how it meets the first and third AOO acceptance criteria for B dilution events. The NRC staff considers that these criteria are met, based on the reasoning described below.

Violating the first criterion would require the addition of heat (e.g., by generating power) or mass (e.g., by operating the emergency core cooling system (ECCS) or the charging system) to the RCS. Heat would not be added, during a B dilution event, if the B dilution does not cause the core to become critical. The addition of mass is evaluated elsewhere in TVA's application, the analyses of the two mass addition events listed in RG 1.70, "NRC Standard Format and Content of Safety Analysis Reports for Nuclear Power Plants: LWR Edition," Revision 2, issued September 1975; the inadvertent operation of the ECCS; and the CVCS malfunction. The results of these analyses show that the first criterion, the RCS pressure safety limit, is met. The licensee's analyses of the feedwater malfunction and the steam generator tube rupture, which add mass to the main steam system, show that the main steam system pressure limit is also satisfied.

Adding mass and/or heat to the RCS can also violate the third criterion, by causing the pressurizer to fill with water and discharging water through the pressurizer power operated relief valves (PORVs). Because the PORVs are not qualified for water relief, they can stick open, and thereby create a more serious event, a loss-of-coolant accident (LOCA), at the top of the pressurizer. This possibility is addressed by the licensee's analyses of the inadvertent operation of the ECCS, and the CVCS malfunction events.

Therefore, the B dilution analysis, and the NRC staff's review of the B dilution analysis are focused upon the second criterion. The NRC staff applied the following measures, from SRP Section 15.4.6, listed below to review B dilution event analyses, with respect to minimum DNBR requirements (i.e., the second criterion).

If terminating the transient requires operator action, the following minimum time intervals must be available between the time an alarm announces an unplanned moderator dilution and the time shutdown margin is lost:

- during refueling: 30 minutes.
- during startup, cold shutdown, hot shutdown, hot standby, and power operation: 15 minutes.

However, TVA applied different benchmarks listed below in its B dilution analyses. These benchmarks appear in TVA's response to RAI 15.0.0-1.b (Enclosure 4 of TVA's letter to the NRC dated December 10, 2010; ADAMS Accession No. ML103480708).

If terminating the transient requires operator action, the following minimum intervals must be available between the initiation of the uncontrolled B dilution event and the time of complete loss of shutdown margin:

- refueling (Mode 6): 30 minutes
- startup and power (Modes 2 and 1): 15 minutes

TVA's B dilution analyses, performed in accordance with these benchmarks, (1) evaluate B dilution occurrences only in Modes 1, 2 and 6, and (2) define the time that is available for operator action to begin at the inception of the dilution, not at the time of an alarm or other indication. Therefore, the NRC staff asked TVA to provide an evaluation or analysis of B dilution events occurring in Modes 3, 4, and 5, showing that there is adequate time available, for remedial action, following receipt of a qualified alarm.

TVA responded by letter dated July 29, 2011 (ADAMS Accession No. ML11215A132), and stated that, "The Watts Bar units were originally licensed to Regulatory Guide (RG) 1.70, Revision 0 and 1 which required explicit Boron Dilution calculations in Modes 1, 2, and 6. Subsequent revisions to RG 1.70 have added requirements to consider boron dilutions in all six operating modes." TVA's assertion that the WBN units were originally licensed to Revision 0 or Revision 1 to RG 1.70, contradicts Chapter 15.0 of the WBN Unit 1 updated FSAR, which states, "This chapter addresses the accident conditions listed in Table 15-1 of the NRC Standard Format and Content Guide, Regulatory Guide 1.70, Revision 2, which apply to WBN." Chapter 15.0 of the WBN Unit 2 FSAR contains the same statement. The NRC staff considers RG 1.70, Revision 2, to be in the licensing basis of WBN Unit 1. Because the Commission has directed that the NRC staff to use the WBN Unit 1 licensing basis as the reference for its review of WBN Unit 2, and because the WBN Unit 2 FSAR references Revision 2 to RG 1.70, the NRC staff considers that Revision 2 to RG 1.70 to be part of the WBN Unit 2 licensing basis. The NRC issued SRP Section 15.4.6, which calls for analysis of the B dilution event in all modes of operation, in November 1975, and issued Revision 2 to RG 1.70 in September 1975. The agency published both documents about a year before the WBN units were docketed on September 15, 1976.

In its RAI response, TVA quoted the following excerpt from Generic Letter (GL) 85-05, "Inadvertent Boron Dilution Events," dated January 31, 1985, "the consequences are not severe enough to jeopardize the health and safety of the public and do not warrant backfitting requirements for boron dilution events at operating reactors." In 1985, WBN Unit 1 was not an operating plant. WBN Unit 1 did not begin commercial operation until February 7, 1996, more than 11 years after GL 85-05 was issued, and 20 years after its license application was docketed. The NRC staff does not consider its request for analyses of the B dilution event in Modes 3, 4, and 5, to be a backfit requirement, since WBN Unit 2 is not an operating plant.

In its letter dated January 12, 2012 (ADAMS Accession No. ML12108A315), TVA supplied analyses of the B dilution event, addressing all modes of plant operation; indicating the time that would be available for operator action following receipt of a qualified alarm.

Table 15.2.4-1 summarizes the B dilution analysis results.

Table 15.2.4-1 B Dilution Results

	Mode 1		Mode 2	Mode 3	Mode 4	Mode 5	Mode 6
	Power		Startup	Hot Standby	Hot Shutdown	Cold Shutdown	Refueling
	With Rod Control	Without Rod Control					
Dilution flow rate (gpm)	235	235	235	160	160	160	0^a
Alarm or trip	Low rod insertion limit	OTΔT Rx trip	High source range flux^b	High VCT level^c	High VCT level^c	High VCT level^c	N/A
Time of alarm/trip (s)	Varies	78	0	820	820	820	N/A
Time to criticality (s)	2,057	2,057	1,584	3,563	3,552	2,186	N/A
Time available (min)	>15	33	26	46	46	23	N/A
Time at 235 gpm^d	N/A	N/A	N/A	31	31	16	N/A

Notes:
a. In Mode 6, B dilution cannot occur due to administrative controls.
b. In Modes 3, 4, and 5; below P-6, and 10^4 counts/sec: high flux at shutdown alarm (setting is automatically reduced as the count rate drops)
c. TVA has committed to add a high volume control tank (VCT) level alarm to the main control room enunciator system.
d. Assuming that the charging flow is greater than 160 gpm (higher than the high flow alarm setpoint) is very conservative, and shows that there are at least 15 minutes available, even at this flow rate.

The table indicates that there is enough time available to prevent the loss of shutdown margin during a B dilution, for Modes 1, 2 and 6. Therefore the SAFDLs will not be exceeded.

<u>Mode 1</u>

At power (Mode 1), the analysis assumed a conservatively high dilution flow of 235 gpm, delivered by three charging pumps. In fact, the third pump, a positive displacement charging pump, was abandoned by TVA; but was nevertheless assumed to be operating for the purposes of the analysis. A low RCS water volume, 8,451 cubic feet, which corresponds to the active RCS volume, minus the volumes of the pressurizer and the reactor vessel upper head, was assumed to be available for dilution. A high initial boron concentration (1,500 ppm) was assumed. This corresponds to the critical concentration at hot full power, with rods inserted to their insertion limits, and without xenon. After the reactor trip, the critical boron concentration was assumed to be 1,250 ppm, which corresponds to hot zero power, with all but the most reactive rod inserted, and without xenon. Therefore, a dilution of just 250 ppm would be enough to cause the core to return to critical.

The analysis considered automatic and manual rod control modes. In automatic control mode, the rods would be inserted to compensate for the power and temperature increase caused by the B dilution, and this would decrease the available shutdown margin. The operator would be alerted by the rod insertion limit alarms that a B dilution is occurring. The analysis results indicate there are more than 15 minutes available for operator action from the time of LOW-LOW rod insertion limit alarm to the loss of available shutdown margin.

In manual rod control mode, the increase in power and temperature would lead to an automatic reactor trip on over-temperature delta-T (ΔT). Following the trip, the operator would have more than 15 minutes to terminate the B dilution before losing shutdown margin.

<u>Mode 2</u>

During startup operations (Mode 2), the reactor will be tripped by the high source range flux trip signal. The trip signal has only two channels (sensors). It is not credited in the rod withdrawal from subcritical event analysis, because a rod withdrawal might go undetected if it occurs in the region of the core that is closest to the sensor that is assumed to have failed. However, a B dilution will cause a relatively uniform reactivity excursion. In this case, the source range high flux signal would still be generated, despite a failed sensor.

The results indicate that 26 minutes would be available after the trip before shutdown margin would be lost even at the conservatively high charging flow rate of 235 gpm.

<u>Mode 3</u>

At hot standby (Mode 3), the TVA's B dilution analysis is based on a charging flow rate of 160 gpm. There is a main control panel enunciator alarm on high charging flow, set to 158 gpm (including uncertainty). A higher flow rate will result in an immediate alarm on high charging flow. The credited alarm is high VCT level.

The results indicate that 46 minutes would be available after the VCT high level alarm. For comparison to cases that were evaluated in higher plant modes, the NRC staff assumed a very high (conservative) charging flow rate of 235 gpm. At this rate, with the high flow alarm sounding, the available time would be reduced to 31 minutes.

Mode 4

The same conditions would apply to hot shutdown (Mode 4); 46 minutes would be available for operator action after the VCT high level alarm. Assuming a conservatively high charging flow rate of 235 gpm, then the available time would be reduced to 31 minutes.

Mode 5

Charging 160 gpm, at cold shutdown (Mode 5), would lead to a high VCT level alarm at 820 seconds. Shutdown margin would be lost 23 minutes later. Assuming the conservatively high charging flow rate of 235 gpm would reduce the available time to 16 minutes. This result still meets the 15 minute acceptance criterion.

Mode 6

During refueling (Mode 6), certain valves are closed, in accordance with administrative controls, which block the flow of unborated makeup water to the RCS.

The results of the B dilution analysis, for all plant operational modes, indicate that there would be adequate time for operator action to terminate the boron dilution before shutdown margin is lost. Therefore, the NRC staff concludes that manual action can prevent the core from becoming critical, because the available time interval, beginning at the receipt of an alarm, exceeds the minimum time requirement in the B dilution acceptance criteria specified in the SRP.

Conclusion

The NRC staff reviewed TVA's analyses of the decrease in boron concentration in the reactor coolant due to a CVCS malfunction and concludes that TVA performed the analyses using acceptable analytical models. The NRC staff further concludes that TVA has demonstrated that the reactor protection and safety systems, and operator action will ensure that the SAFDLs and the RCPB pressure limits will not be exceeded as a result of this event, for all modes of plant operation. Based on this, the NRC staff concludes that the plant will meet the requirements in GDCs 10, 15, and 26 of Appendix A to 10 CFR Part 50 in the event of a decrease in boron concentration in the reactor coolant due to a CVCS malfunction. Therefore, **Open Item 132 is closed**.

APPENDIX A

CHRONOLOGY OF RADIOLOGICAL REVIEW OF
WATTS BAR NUCLEAR PLANT, UNIT 2, OPERATING LICENSE REVIEW

Public correspondence exchanged between the NRC and TVA during the review of the operating license application for Watts Bar Nuclear Plant (WBN), Units 1 and 2, is available through the NRC's Agencywide Documents Access and Management System (ADAMS) or the Public Document Room (PDR). This correspondence includes that occurring subsequent to TVA's letter notifying the NRC of its decision to reactivate construction of WBN Unit 2, which had been in a deferred status under the Commission's Policy Statement on Deferred Plants.

Web-based ADAMS (WBA) is the latest interface to ADAMS. This search engine enables searching the ADAMS repository of official agency records (Publicly Available Records System (PARS) and Public Legacy libraries) for publicly available regulatory guides, NUREG-series reports, inspection reports, Commission documents, correspondence, and other regulatory and technical documents written by NRC staff, contractors, and licensees. WBA permits full-text searching and enables users to view document images, download files, and print locally. New documents become accessible on the day they are published, and are released periodically throughout the day. ADAMS documents are provided in Adobe Portable Document Format (PDF).

The NRC PDR reference staff is available to assist with ADAMS. Contact information for the PDR staff is on the NRC Web site at http://www.nrc.gov/reading-rm/contact-pdr.html.

APPENDIX E

PRINCIPAL CONTRIBUTORS TO SSER 26

D. Allsopp, NRR/DIRS/IOLB
R. Alvarado, NRR/DE/EICB
L. Brown, NRR/DRA/AADB
N. Carte, NRR/DE/EICB
F. Lyon, NRR/DORL/LPWB
P. Milano, NRR/DORL/LPWB
S. Miranda, NRR/DSS/SRXB
T. Mossman, NRR/DE/EICB
J. Parillo, NRR/DRA/AADB
J. Poehler, NRR/DE/EVIB
J. Poole, NRR/DORL/LPL2-2
C. Moulton, NRR/DRA/AFPB
D. Frumkin, NRR/DRA/AFPB
C. Cooper, NRR/DRA/AFPB
B. Metzger, NRR/DRA/AFPB
B. Litket, NRR/DRA/AFPB

APPENDIX FF

FIRE PROTECTION PROGRAM SAFETY EVALUATION

WATTS BAR NUCLEAR PLANT, UNITS 1 AND 2

1.0   INTRODUCTION

The Tennessee Valley Authority (TVA) is the licensee for Watts Bar Nuclear Plant (WBN) Unit 1 and is the applicant for an operating license for WBN Unit 2. TVA submitted the As-Designed Fire Protection Report (FPR) for WBN Units 1 and 2 to the U.S. Nuclear Regulatory Commission (NRC) by letter dated December 18, 2010, as revised and supplemented by letters dated December 20, 2010; January 14, March 16 and 31, May 6, 18, and 26, June 7 and 17, July 1 and 22, August 5 and 15, September 30, October 28, November 21 and 30, 2011; March 13, April 12, 17, and 26, May 9 and 30, June 7 and 27, July 19, September 13, December 20, 2012; February 7 and 28, and March 13, 2013.

In the FPR, TVA stated that, "the purpose of the Fire Protection Report (FPR) is to consolidate a sufficiently detailed summary of the WBN regulatory required Fire Protection Program into a single document and to reflect the design as-constructed at the time of fuel load." The FPR describes the operational phase of the fire protection program. Accordingly, the NRC staff reviewed the entire fire protection program (except as noted otherwise) using the agency's fire protection requirements and review guidance. Because WBN consists of two units of identical design, this evaluation applies to the fire protection program for both WBN Unit1 and WBN Unit 2 (except as noted otherwise).

The NRC staff's review did not include Section 7, "Unit 1 Operator Manual Actions [OMAs]," of Part VII of the FPR. The NRC's approval of the WBN Unit 1 OMAs is documented in Supplemental Safety Evaluation Report (SSER) 18, NUREG-0847, "Safety Evaluation Report Related to the Operation of Watts Bar Nuclear Plant Units 1 and 2," dated October 1995.

TVA's fire protection program is required to comply with the following:

- General Design Criterion (GDC) 3, "Fire Protection," of Appendix A, "General Design Criteria for Nuclear Power Plants," to Title 10 of the *Code of Federal Regulations* (10 CFR) Part 50, "Domestic Licensing of Production and Utilization Facilities"

- 10 CFR 50.48, "Fire Protection," paragraph (a),

In addition to these requirements, TVA commited in the FPR that its fire protection program has been developed to comply with, and is based on, the requirements of:

- Sections III.G, III.J, III.L, and III.O of Appendix R, "Fire Protection Program for Nuclear Power Facilities Operating Prior to January 1, 1979," to 10 CFR Part 50

- Appendix A to Auxiliary Power Conversion Systems Branch (APCSB) Branch Technical Position (BTP) 9.5-1, "Guidelines for Fire Protection for Nuclear Power Plants Docketed Prior to July 1, 1976."

In the FPR, TVA additionally stated that the applicable guidelines used as the basis for the plan included, in part, the following:

- NRC letter entitled, "Nuclear Plant Fire Protection Functional Responsibilities, Administrative Controls and Quality Assurance," dated June 20, 1977

- Generic Letter (GL) 81-12, "Fire Protection Rule (45 FR 76602, November 19, 1980)," dated February 20, 1981, and its associated clarification letter, dated March 22, 1982;

- GL 82-21, "Technical Specifications for Fire Protection Audits," dated October 6, 1982;

- GL 83-33, "NRC Positions on Certain Requirements of Appendix R to 10 CFR 50," dated October 19 1983;

- GL 86-10, "Implementation of Fire Protection Requirements," dated April 24, 1986;

- GL 88-12, "Removal of Fire Protection Requirements from Technical Specifications," dated August 2, 1988.

The following NRC guidance was used for specific topics:

- NUREG-1852, "Demonstrating the Feasibility and Reliability of Operator Manual Actions in Response to Fire," issued October 2007, for WBN Unit 2 OMA evaluations

- NRC Regulatory Guide (RG) 1.189, "Fire Protection for Operating Nuclear Power Plants," Revision 0, issued April 2001, for extension of the "annual" fire protection audit interval

- NRC RG 1.189, "Fire Protection for Nuclear Power Plants," Revision 2, issued October 2009, for OMA and multiple spurious operation (MSO) evaluations.

In Staff Requirements Memorandum SECY-07-0096, "Possible Reactivation of Construction and Licensing Activities for the Watts Bar Nuclear Plant Unit 2," dated July 25, 2007, the Commission directed the NRC staff to use the existing WBN Unit 1 licensing basis as the reference basis for the WBN Unit 2 review. To that end, where applicable, the NRC staff used the WBN Unit 1 approvals, as documented in SSER 18, issued October 1995, and SSER19, issued November 1995, to NUREG-0847, as the basis for its approvals in this evaluation, instead of the agency's current guidance. The NRC staff used the agency's current guidance as the basis for approval for the WBN Unit 2 OMAs, associated circuits, MSO, fire water system design demand, the auxiliary control room (ACR), and radiant energy shields (RES).

The NRC staff met with TVA on January 19, February 3 and 15, March 29, April 22, May 12, June 30, July 12 and 28, August 31, November 16, and December 21, 2011, and February 2, 2012, to discuss technical issues related to WBN's fire protection program and its implementation. The NRC staff also conducted an audit at WBN from October 25-27, 2011, which it documented in a report dated December 20, 2011 (Agencywide Documents Access and Management System (ADAMS) Accession No. ML113500239).

Unless otherwise noted, all information cited in this evaluation is from the WBN FPR dated March 13, 2013 (ADAMS Accession No. ML130840169).

2.0   FIRE PROTECTION PROGRAM

2.1   Purpose and Scope

In FPR Part I, Section 2.0, "Purpose," TVA stated that the purpose of the FPR is to provide a detailed summary of the WBN fire protection program in a single document. The FPR is thus the "fire protection plan" document that is required by 10 CFR 50.48(a). Section 9.5.1 of the WBN Final Safety Analysis Report (FSAR) incorporates the FPR by reference. In FPR Part I, TVA states that it will be updated in conjunction with the FSAR.

The regulation at 10 CFR 50.48(a)(1) requires that licensees have a fire protection plan that satisfies General Design Criterion 3 in Appendix A to 10 CFR 50. The plan must do the following:

(i)   Describe the overall fire protection program for the facility;

(ii)   Identify the various positions within the licensee's organization that are responsible for the program;

(iii)   State the authorities that are delegated to each of these positions to implement those responsibilities; and

(iv)   Outline the plans for fire protection, fire detection and suppression capability, and limitation of fire damage.

TVA's plan provided information on Item (i) above in FPR Part II, Section 9, "Emergency Response," Section 10, "Control of Combustibles," and Section 11, "Control of Ignition Sources." TVA's plan provided information on Item (ii) above in FPR Part II, Section 7, "Fire Protection Organization/Programs," and Section 14, "Fire Protection Systems and Features Operating Requirements," and in FPR Part VI. TVA's plan provided information on Item (iii) above in FPR Part II, Section 7, "Fire Protection Organization/Programs," and in FPR Parts III, IV, V, and VI. TVA's plan provided information on Item (iv) above in FPR Part II, Section 12, "Description of Fire Protection Systems and Features." Items (i) through (iii) are evaluated in Section 2.0 of this safety evaluation. Item (iv) is evaluated in Sections 2.0 through 5.0 of this safety evaluation.

The regulation at 10 CFR 50.48(a)(2) requires that the plan must describe specific features necessary to implement the program described in 10 CFR 50.48 (a)(1), such as the following:

(i)   Administrative controls and personnel requirements for fire prevention and manual fire suppression activities;

(ii)   Automatic and manually operated fire detection and suppression systems; and

(iii)   The means to limit fire damage to structures, systems, or components (SSCs) important to safety so that the capability to shut down the plant safely is ensured.

TVA's plan provided information on Item (i) above in FPR Part II, Section 9, "Emergency Response," Section 10, "Control of Combustibles," Section 11, "Control of Ignition Sources," and

Section 13, "Fire Protection System Impairments." TVA's plan provided information on Item (ii) above in FPR Part II, Section 12, "Description of Fire Protection Systems and Features." TVA's plan provided information on Item (iii) above in FPR Part II, Section 12, "Description of Fire Protection Systems and Features," and in FPR Parts III, IV, V, and VI. Item (i) is evaluated in Section 2.0 of this safety evaluation. Item (ii) is evaluated in Section 4.0 of this safety evaluation. Item (iii) is evaluated in Sections 3.0 and 5.0 of this safety evaluation.

The regulation at 10 CFR 50.48(a)(3) requires the licensee to retain the fire protection plan and each change to the plan as a record until the reactor license is terminated. In FPR Part I, Section 2 "Purpose," TVA stated that the FPR will be updated in conjunction with updates to the WBN FSAR. The NRC staff concludes that this an acceptable method of retaining plan records, because the FSAR is maintained and updated in accordance with 10 CFR 50.59, "Changes, Tests, and Experiments," and 10 CFR 50.71(e), respectively, which have similar retention requirements and therefore meets the requirements of 10 CFR 50.48(a)(3).

The information below describes how TVA organized its FPR.

FPR Part I is an introduction to the FPR and contains a summary table of fire protection features throughout the plant. FPR Part II of the FPR contains the overall fire protection plan. The fire protection plan describes (1) the WBN fire protection organization, (2) plant fire protection features, (3) the plant's fire prevention program, (4) the plant's emergency response organization, (5) plant operating requirements for fire protection features and systems, and (6) the testing and inspection requirements for these plant fire protection features. An overview of the post-fire safe shutdown (FSSD) is contained within FPR Part III. FPR Part IV of the FPR discusses alternate shutdown. FPR Part V describes OMAs and repairs. In FPR Part VI, the FPR summarizes the fire hazards analysis (FHA) for each fire area by describing the physical characteristics of the fire area, combustible loadings and anticipated fire severity, and fire suppression and detection capability available in each plant area. In FPR Part VI, TVA also describes how the plant would achieve post-FSSD if a serious fire occurred in the fire area. FPR Part VII documents deviations from regulatory criteria and guidance documents and presents engineering evaluations related to the adequacy of specific fire protection features. FPR Parts VIII and IX describe conformance with the guidelines in Appendix A to BTP (APCSB) 9.5-1 and in Sections III.G, III.J, III.L, and III.O of Appendix R to 10 CFR Part 50, respectively. FPR Part X contains a discussion of TVA's compliance with National Fire Protection Association (NFPA) codes.

The FPR describes the measures that are established at WBN to implement a defense-in-depth fire protection program in plant areas important to safety. The objective of these measures is to: (1) prevent fires from starting; (2) detect rapidly, control, and extinguish promptly those fires that do occur; and (3) provide protection for SSCs important to safety so that a fire that is not promptly extinguished by the fire suppression activities will not prevent the safe shutdown of the plant.

2.2   Fire Protection Organization

As described in FPR Part II, Section 7, TVA's fire protection organization consists of corporate management oversight and an onsite plant implementation organization. Responsible TVA corporate managers include the Senior Vice President, the Engineering Vice President, and the Site Vice President. The onsite implementation organization includes the Plant Manager, the Operations Manager, the Operations Support Supervisor, the Fire Protection Supervisor, and the Site Engineering Manager. The NRC staff reviewed the responsibilities and authorities of each position responsible for the fire protection program, as described in FPR Part II, Sections 7.1 through 7.6, and concluded that there is reasonable assurance that the key responsibilities for implementing the fire protection program at WBN have been delegated to appropriate positions within TVA's organization, and that the authorities delegated to each position to implement these responsibilities are appropriate.

Based on its review of the FPR, the NRC staff concludes that TVA's fire protection organization does not take any exceptions to Position A.1 of Appendix A to BTP (APCSB) 9.5-1, therefore, is acceptable.

2.3   Fire Protection Quality Assurance Program

FPR Part II, Section 6.0, contains TVA's description of the quality assurance (QA) program for fire protection at WBN. TVA stated that it used the guidance established by Appendix A to BTP (APCSB) 9.5-1 and the NRC's letter dated June 20, 1977, "Nuclear Plant Fire Protection Functional Responsibilities, Administrative Controls, and Quality Assurance," to develop a QA program for fire protection features that protects post-FSSD capability and safety-related SSCs. The FPR states that the WBN fire protection QA program uses the applicable parts of TVA-NQA-PLN89-A, "Tennessee Valley Authority Nuclear Quality Assurance Plan."

TVA implemented a program that performs independent audits and inspections of the WBN fire protection program. TVA stated that its program is based on the guidance in GL 82-21. The FPR states that TVA's Nuclear Assurance organization is responsible for conducting the fire protection-related audits.

In TVA's letter dated May 6, 2011 (ADAMS Accession No. ML11129A158), in response to the NRC staff's request for additional information (RAI) FPR II-26, TVA stated that the frequency of the GL 82-21 annual fire protection audit has been changed to 24 months. TVA stated in its letter dated August 28, 2002, (ADAMS Accession No. ML022460173) that the plant implemented this change using a performance-based schedule. In TVA's letter dated September 30, 2011 (ADAMS Accession No. ML13060A225), in response to the NRC's question RAI FPR II-26.1, TVA stated that the change is being monitored on a fleet-wide basis, and that deficiencies found during the biennial audits would result in increasing the frequency of the audits. The NRC staff concludes that this is consistent with Position 1.7.10.1 of Revision 0 to RG 1.189, and, therefore, is acceptable.

Based on its review of the information submitted by TVA, the NRC staff concludes that TVA's fire protection QA program does not take any exceptions to Position C of Appendix A to BTP (APCSB) 9.5-1 and, therefore, is acceptable.

2.4   Fire Protection Administrative and Technical Controls

2.4.1   Fire Protection Program Changes, Review and Approval

TVA stated in FPR Part I, Section 2.0, that "the Fire Protection Report has been developed in accordance with the guidelines of NRC Generic Letter 86-10...and NRC Generic Letter 88-12...." TVA has elected to follow the guidance in GL 88-12 and incorporate the standard fire protection license condition as listed in GL 86-10. In addition to including, by reference, the NRC safety evaluations which approved the plant fire protection program, this license condition allows TVA to make changes to the approved program without prior approval of the NRC if those changes would not adversely affect the plant's ability to achieve and maintain safe shutdown in the event of a fire.

Based on its review of the information submitted by TVA, the NRC staff concludes that no exceptions were taken to the positions in GL 88-12, and it is therefore, acceptable.

2.4.2   Fire Protection Administrative Controls

2.4.2.1   Control of Combustibles

FPR Part II, Section 10.0, describes TVA's program to control combustibles. The WBN combustible control program objectives are to (1) provide instruction and guidelines during general employee training on the application and use of combustible materials at WBN, (2) control the application and use of chemicals, (3) perform periodic plant housekeeping inspections and have housekeeping tours by management and the onsite fire protection organization, (4) control in situ combustibles through the design/modification review and installation process, and (5) control transient combustibles through the implementation of administrative controls.

TVA stated that it has established a plantwide administrative procedure to control transient combustibles. Implementation of this procedure will establish administrative controls for the handling of combustible materials such as fire-retardant wood, paper, plastic, and flammable and combustible gases and liquids. In addition, through its combustible control program, TVA has established combustible control zones in the plant. TVA considers these zones to be subdivisions of fire areas and to limit fire spread by providing open space free of transient combustibles between redundant FSSD equipment or cables. Transient combustibles may not be stored in these zones unless an adequate fire protection engineering evaluation or compensatory measures, or both, are implemented.

Based on its review of the information submitted by TVA, the NRC staff concludes that TVA's program to control combustibles does not take any exceptions to Positions B.2 and B.3.c of Appendix A to BTP (APCSB) 9.5-1 and, therefore, is acceptable.

2.4.2.2   Control of Ignition Sources

TVA has established a program for controlling ignition sources such as welding, cutting, grinding, and the use of open flame. TVA's program specifies that the issuance of "hot work" permits be reviewed and approved based on plant conditions and a prior inspection of the proposed work area. The ignition source on a hot work permit is valid for only one job. Before the start of work, the work area is made "fire safe." In addition, TVA's program will establish a hot work fire watch for all ignition source work activities that are performed in safety-related and safe-shutdown areas of the plant. These fire watches, in addition to performing their duties during the hot work activities, will remain in the area for a minimum of 30 minutes after the work has been completed to ensure that potential residual ignition conditions do not exist.

Based on its review of the information submitted by TVA, the NRC staff concludes that TVA's program to control ignition sources does not take any exceptions to Positions B.3.a and B.3.b of Appendix A to BTP (APCSB) 9.5-1 and, therefore, is acceptable.

2.4.3 Compensatory Measures

Compensatory measures described in FPR Part II are used to compensate for degraded or nonfunctional fire protection systems or features. Primarily, these compensatory measures take the form of both roving and continuous fire watches.

FPR Part II, Section 13.B states, "A roving fire watch consists of a trained individual in an affected location at 60 minute intervals with a 15 minute margin to accommodate and handle unforeseen circumstances and to report and/or resolve potential fire hazards in a location. Roving fire watches are required as a compensatory action in all modes of plant operation (i.e., Modes 1 through 6 or core empty)." The NRC staff concludes that this takes no exceptions to Positions B.3 and B.5.a of Appendix A to BTP (APCSB) 9.5-1 and, therefore, is acceptable.

As described in FPR Part II, Section 13.A, a continuous fire watch possesses the following attributes: (1) the trained person performing the fire watch must be in the fire area at all times; (2) the fire area must not contain any impediment to restrict the movements of the fire watch; and (3) each compartment within the fire area must be patrolled at least once every 15 minutes with a margin of 5 minutes. The NRC staff concludes that this takes no exceptions to Positions B.3 and B.5.a of Appendix A to BTP (APCSB) 9.5-1 and, therefore, is acceptable.

In the FPR, TVA identified specific exceptions to the above requirements for roving and continuous fire watches. In FPR Section 13.0, TVA identified continuous fire watch routes in more than one fire area that it classifies as exceptions to a continuous fire watch remaining within one fire area. As a basis for acceptability, TVA identified the following characteristics: (1) one or more rooms in different fire areas whose proximity to one another and their limited size warrant the combining of them into one continuous fire watch route, (2) a time study that confirms the route can be covered in 15 minutes without putting undue exertion on the person performing the fire watch, and (3) in each instance, these routes require the Fire Protection Supervisor's approval to ensure that the conditions that formed a basis for the time study have not changed in such a manner as to invalidate the time study. In the event that the automatic suppression or detection systems in the above areas cannot be restored within the time specified by FPR Part II, Section 14.0, TVA stated that the continuous fire watch patrols would not be allowed to include more than one fire area. Based on the submitted information, the NRC staff concludes that this takes no exceptions to Positions B.3 and B.5.a of Appendix A to BTP (APCSB) 9.5-1 and, therefore, is acceptable.

The WBN FPR states that continuous fire watches are only required when the affected unit is in Mode 1 (power operation) through Mode 4 (hot shutdown). In FPR Part II, TVA stated that, when one unit is in Modes 5, 6, or core empty, locations where a continuous fire watch would be required may be combined and patrolled by a roving fire watch when approved by the Fire Protection Supervisor, if a fire in those locations could not affect the other unit, if it is in Modes 1 through 4. Based on the submitted information, the NRC staff concludes that this takes no exceptions to Positions B.3 and B.5.a of Appendix A to BTP (APCSB) 9.5-1 and, therefore, is acceptable.

In addition, in the FPR Part II, Section 13.0, TVA identified other alternative compensatory measures that may be used at WBN in lieu of the above standard compensatory measures. In all cases, in which an alternative compensatory measure is used for a degraded or nonfunctional fire protection feature, TVA stated that it will perform an evaluation that demonstrates technical equivalency to the standard compensatory measure identified in FPR Part II, Section 14.0. TVA described the following alternatives that may be considered when supported by an appropriate technical evaluation: (1) providing additional or alternative fire protection equipment, (2) installing temporary or portable fire detection systems in conjunction with an hourly roving fire watch, (3) installing closed circuit television cameras and monitors in areas when special circumstances, such as personal safety or as-low-as reasonably-achievable (ALARA; radiological) concerns, preclude the use of a human fire watch in the area, and (4) taking credit in continuously manned areas for the constant manning in lieu of establishing either continuous or roving compensatory fire watches when the responsible individuals accept this responsibility. Based on its review of the information submitted by TVA, the NRC staff concludes that these alternatives take no exceptions to Positions B.3 and B.5.a of Appendix A to BTP (APCSB) 9.5-1 and, therefore, are acceptable.

2.4.4   Fire Protection Technical Controls

In FPR Part II, Section 14, TVA established operability requirements for the following fire protection features: (1) fire detection instrumentation, (2) water supply, (3) water-based fire suppression systems, (4) carbon dioxide (CO_2) suppression systems, (4) fire detection supervisory equipment, (6) fire hose stations and associated pre-action control valves, (7) fire hydrants, (8) fire-rated assemblies, and (9) emergency battery lighting units.

Based on its review of the information submitted by TVA, the NRC staff concludes that TVA's operability requirement program for plant fire protection features does not take any exceptions to Positions B.1, B.3, and B.5 of Appendix A to BTP (APCSB) 9.5-1, and, therefore, is acceptable.

GL 88-12 provides guidance for removing fire protection limiting conditions for operation and surveillance requirements associated with fire detection systems, fire suppression systems, fire barriers, and administrative controls that address fire brigade NRC staffing from the plant's technical specifications (TSs) and incorporating this information into the FSAR. In addition, GL 88-12 refers to GL 81-12, which requested licensees to provide TSs for equipment used for safe shutdown capability that is not currently covered by existing TSs. In its fire protection plan, TVA confirmed that the plant equipment used to achieve and maintain post-FSSD from either inside or outside the main control room (MCR) is included in either the plant TSs or the FPR.

Table 14.10, "Fire Safe Shutdown Equipment," of FPR Part II, Section 14, lists the FSSD equipment not included in the plant's TSs. TVA established testing and inspection requirements which assist in evaluating the operability of the non-TS-related FSSD equipment and instrumentation. In FPR Part II, Section 14.0, TVA established the requirements with this equipment or instrumentation inoperable. TVA requires, with one or more of the required items of equipment listed in Table 14.10 inoperable (or a breaker or valve not in its safe shutdown position), that the plant restore the equipment to the operable status within 30 days, or that it either: (1) place the equipment in the condition required for FSSD, (2) provide a backup means of instrumentation monitoring, (3) provide an alternative means of achieving post-FSSD (along with an evaluation justifying the alternative), or (4) be in Mode 3 within 6 hours and Mode 4 within the following 12 hours.

Based on the information provided in FPR Part II, the NRC staff concludes that TVA's removal of fire protection features from the plant's TSs and relocation to the FPR as operating requirements is consistent with the guidance in GL 88-12 and GL 81-12, and, therefore, is acceptable.

In addition, in FPR Part II, Section 14, TVA established testing and inspection requirements for the following fire protection features: (1) fire detection instrumentation, (2) water supply, (3) water-based fire suppression systems, (4) CO_2 suppression systems, (5) fire detection supervisory equipment, (6) fire hose stations and associated pre-action control valves, (7) fire hydrants, (8) fire-rated assemblies, (9) emergency battery lighting units, and (10) the FSSD equipment identified in Table 14.10.

Based on its review of the information submitted by TVA, the NRC staff concludes that TVA's surveillance and test program for plant fire protection features does not take any exceptions to Position B.5 of Appendix A to BTP (APCSB) 9.5-1, and, therefore, is acceptable.

2.5   Fire Brigade and Fire Response

2.5.1   Organization

FPR Part II, Section 9.1, "Fire Brigade NRC Staffing," states that a fire brigade comprising of at least five members will be maintained on site at all times. In the FPR, TVA stated that these five members will consist of the fire brigade leader and four fire brigade members. In addition, neither the shift operations supervisor nor the other members of the operations shift crew needed to perform a safe shutdown of the WBN units will be included in the fire brigade. In addition, the fire brigade will not include any other individuals required for other essential plant functions that may be necessary during a fire emergency.

TVA also stated that an incident commander is available to direct each shift fire brigade in addition to the five-member fire brigade. The incident commander has sufficient knowledge of plant safety systems to understand the effects of fire and fire suppression on safe shutdown capability.

TVA stated that before initial training and annually thereafter its fire brigade program requires each fire brigade member to undergo a medical review and to receive medical approval to perform strenuous physical activities related to fire fighting and to wear special respiratory equipment.

TVA stated that the fire brigade may comprise of less than five members for a period of time not to exceed 2 hours, to accommodate for unexpected conditions such as an unplanned absence, or brigade response to a non-fire emergency.

Based on its review of the information submitted by TVA, the NRC staff concludes that TVA's fire brigade NRC staffing and organization does not take any exceptions to Positions B.4 or B.5 of Appendix A to BTP (APCSB) 9.5-1 and, therefore, is acceptable.

2.5.2   Training

FPR Part II, Section 9.3 "Training and Qualifications," states that TVA's fire brigade training program consists of initial training, recurrent training, and annual fire brigade training.

The initial training program includes: (1) instruction and practical exercises in fire extinguishment and the use of fire-fighting equipment, (2) identification of fire hazards and types of fires that could occur in the plant, (3) identification of the location of fire-fighting equipment in each fire area of the plant, (4) instruction on the proper use of plant fire-fighting equipment, (5) instruction on the proper use of communications, lighting, ventilation, and emergency breathing apparatus, (6) instruction on the toxic characteristics of the products of combustion, and (7) instruction and practical exercises in fighting fires inside buildings and tunnels. In addition to initial training, the program instructs the fire brigade is instructed on fire-fighting procedures and procedure changes, the plant fire-fighting plan, with emphasis on each individual's responsibility, and the latest plant modifications and changes affecting the fire-fighting plans.

The recurrent training consists of classroom instruction meetings held every 3 months. These meetings repeat the initial training subjects over a 2 year period. Each member of the fire brigade is required to attend this training in order to remain qualified. TVA preplans fire brigade drills to establish the objectives, and the fire brigade training instructor or the instructor's designee conducts these drills. The conduct of onsite fire brigade drills are conducted as follows: (1) a minimum of one drill per fire brigade shift will be conducted every 92 days, (2) a minimum of one unannounced drill will be conducted per fire brigade shift per year, and (3) at least one drill per fire brigade shift will be conducted on the backshift. Each fire brigade member is required to attend at least two drills per year.

TVA holds annual training for each fire brigade member. TVA stated that this training provides instruction, under actual fire-fighting conditions, on the proper methods for fighting various types of fires similar in magnitude, complexity, and difficulty to those that could be encountered in the plant. This training includes actual fire extinguishment and the use of fire-fighting equipment under strenuous conditions. TVA stated that if a brigade member misses or does not complete a training session, either annual or quarterly; the member is placed in an ineligible status until the training is completed.

In addition to the annual fire brigade training, TVA holds annual briefings for the local fire departments to ensure their continued understanding of their role in the event of a fire emergency at the site. TVA also holds an annual drill for the local fire department and the plant fire brigade. The local fire department briefings and drills are held for those departments that have active aid agreements with the plant.

Based on its review of the information submitted by TVA, the NRC staff concludes that TVA's fire brigade training program does not take any exceptions to Positions B.5.b and B.5.c of Appendix A to BTP (APCSB) 9.5-1 and, therefore, is acceptable.

2.5.3 Equipment

In the FPR, TVA stated that fire-fighting equipment is provided throughout the plant and is strategically placed near the fire hazards present or anticipated. TVA stated that delays in the fire brigade obtaining fire-fighting equipment are minimized because of the distribution and availability of this equipment throughout the plant. TVA further stated that firefighting equipment may be staged adjacent to, or at the access to, areas/locations to facilitate equipment availability or to address equipment surveillance test concerns relative to life safety and ALARA practices.

The equipment available to the fire brigade includes: (1) motorized fire-fighting apparatus, (2) portable ventilation equipment, (3) fire extinguishers, (4) self-contained breathing apparatus, (5) fire hose, nozzles, and fittings, (6) foam equipment, (7) personal protective equipment, (8) communications equipment, (9) portable lighting, and (10) ladders specifically dedicated for fire-fighting.

Based on its review of the information submitted by TVA, the NRC staff concludes that no exceptions were taken to Position B.5.d of Appendix A to BTP (APCSB) 9.5-1 and therefore, TVA's fire brigade is acceptably equipped.

2.5.4   Fire Emergency Procedures and Pre-Fire Plans

As described in the FPR, TVA's fire emergency procedures and pre-fire plans specify the actions that the individual who discovers a fire must take and the actions that the emergency response organization must consider (e.g., control room operators and the plant fire brigade). These procedures provide different levels of response based on whether actual fire/smoke conditions are reported or whether a fire detection system annunciation occurs. (For example, a single fire detection system zone annunciation in a cross-zoned area will not carry the same level of response as a cross-zone annunciation in the same area).

TVA stated that it has implemented pre-fire plans to provide guidance, depending on the particular circumstances, to aid in firefighting efforts. TVA has developed pre-fire plans to support the fire-fighting activities in plant areas important to safety. Specifically, these plans are developed for safety-related areas, FSSD areas, and areas that present a hazard to safety-related equipment or plant shutdown.

The pre-fire plans provide the following information to the fire brigade: (1) plant equipment in the fire area, (2) access and egress routes to the fire area, (3) fire-fighting strategy and tactics, (4) locations of fire protection features and equipment, (5) special fire, toxic, and radiological hazards in the area, (6) special precautions, and (7) ventilation methodology.

Based on its review of the information submitted by TVA, the NRC staff concludes that TVA's proposed fire brigade preplans and fire emergency procedures do not take any exceptions to either the NRC letter dated June 20, 1977, or 10 CFR 50.48(a)(2) and, therefore, are acceptable.

3.0   GENERAL PLANT FIRE PROTECTION AND SAFE SHUTDOWN FEATURES

3.1   Fire Protection Design

3.1.1 Building and Compartment Fire Barriers

TVA stated that the fire rated assemblies at WBN are part of the passive fire protection features that ensure that one set of redundant FSSD components necessary to achieve and maintain FSSD remains free of fire damage. At WBN, fire-rated assemblies consist of fire barriers, raceway protection, fire doors, fire dampers, and penetration seals.

At WBN, fire areas are defined by rated wall and floor/ceiling assemblies. TVA stated that fire areas are separated by wall and floor/ceiling assemblies that are 2- or 3-hour equivalent fire barriers that are bounded by Underwriters Laboratories (UL), Inc., rated designs. In FPR Part II, Sections 12.10 and 12.10.1, TVA states that the walls that separate buildings and walls between rooms that contain safe shutdown systems are fire-rated assemblies. Rooms within each fire area may be separated from other rooms in the same fire area by regulatory or non-regulatory fire barriers. Where barriers are needed between rooms, TVA stated that only fire rated barriers with a minimum 2-hour rating are relied upon, except for portions of the MCR complex that have 1-hour rated barriers. Sections 6.2.3, 6.2.5, 6.2.6, and 6.2.7 of this evaluation provide NRC staff evaluations of deviations to fire barrier ratings.

In general, the fire barriers comprising compartment walls and floors/ceilings at WBN are constructed of reinforced concrete or concrete block. The reinforced concrete fire barriers and concrete block barriers are at least 8 inches thick. TVA's evaluation of reinforced concrete barriers used information from Section 6, Chapter 5 of the NFPA *Fire Protection Handbook*, 17th Edition (hearafter referred to as the Handbook). This section of the Handbook correlates fire rating and thickness of reinforced concrete. On this basis, Figure 6-5G in the Handbook shows that 6 inches of reinforced concrete has a fire resistance of approximately 4 hours. The concrete block barriers are only used when barriers are required to have a fire rating of 2-hours or less. TVA's evaluation of these fire barrier designs concludes these are similar to UL listed concrete block barrier designs (Designs Nos. U904, U905, U906, and U907) which are 2- to 4-hour fire-rated.

Based on its review of the information submitted by TVA, the NRC staff concludes that, with the exception of items evaluated elsewhere in this evaluation, TVA's proposed technical basis for the fire resistive capability of fire area boundaries offers an equivalent level of fire safety to that of Position D.1.j of Appendix A to BTP (APCSB) 9.5-1 and, therefore, is acceptable.

3.1.2 Fire Barriers Used To Separate Redundant Safe Shutdown Functions within the Same Fire Area

Cable raceways that require separation due to redundant trains located in the same fire area, excluding primary containment and secondary containment (the annulus), are separated by either 1- or 3-hour fire rated barrier systems. TVA uses a 1-hour fire rated barrier system if automatic detection and automatic suppression are installed in the areas and uses a 3-hour fire rated barrier system if automatic suppression is not installed in the area. Cable raceways that require separation due to redundant trains inside the reactor building, which includes primary containment (WBN Unit 1 only) and secondary containment (i.e., the annulus) (both units), rely on RESs or automatic detection and suppression to provide separation. RESs are addressed in Section 6.1.2 of this evaluation.

In FPR Part II, Section 12.10.2, TVA stated that the 1- and 3-hour fire rated barriers were tested in accordance with the guidance in Supplement 1, "Fire Endurance Test Acceptance Criteria for Fire Barrier Systems Used to Separate Redundant Safe Shutdown Trains within the Same Fire Area," to GL 86-10. This guidance includes test parameters, thermocouple placement, conduit and cable tray configurations, hose stream tests, and ampacity derating. TVA also evaluated fire barriers for seismic considerations. Configurations of raceway fire barriers that are not consistent with the testing have been evaluated to ensure that untested configurations are bounded by tested configurations. TVA has procedural controls for evaluating field changes to designed configurations. TVA stated that personnel who perform such field changes are to be cognizant of the important parameters.

Based on its review of the submitted information, the NRC staff concludes that TVA's use of the guidance in Supplement 1 to GL 86-10, with the consideration of seismic events, bounding of untested configurations, and procedures to control field changes, offers an equivalent level of fire safety to that of Position D.3 of Appendix A to BTP APCSB 9.5-1 and, therefore, is acceptable.

3.1.3   Equipment Hatches and Stairwells

TVA stated that at WBN equipment hatches in the floor or fire barriers in the ceiling can be categorized as follows:

- precast concrete plugs
- steel covers
- open hatches and stairwells

TVA stated that the precast concrete plugs are associated with radiation shielding and, as fire barriers, are equivalent to the floor or ceiling fire barrier in which they are located. TVA stated that the steel covers are of substantial construction and that they provide an effective barrier to prevent fire from propagating from one side of the barrier to the other. In addition, because the covers are not fire rated, they are either provided with a draft stop and water curtain around them or redundant safe shutdown components on either side have been separated from each other by a cumulative horizontal distance of 20 feet or more. In either case, automatic fire suppression and detection are provided on both sides of the equipment hatch cover.

FPR Part VII, Section 2.6.4, summarizes TVA's evaluation for the deviation of the non-rated equipment hatches separating the control building and turbine building. Section 6.2.7.4 of this evaluation provides the NRC staff evaluation of this deviation.

TVA stated that, in areas in which open hatches and stairwells are located, redundant shutdown trains are either separated by at least 20 feet horizontally, one train has been protected by a 1-hour fire barrier, or a water curtain has been installed around the opening. In any case, fire detection and automatic suppression systems are located on both sides of the openings. Further, TVA stated that the only exception to this arrangement is in the refueling area of the auxiliary building.

FPR Part VII, Section 2.6.3, summarizes TVA's evaluation of the deviation for non-rated open hatches and stairwells that do not fully meet the NRC staff guidance. Section 6.2.7.3 of this evaluation provides the NRC staff evaluation of this deviation.

FPR Part VII, Section 4.5, summarizes TVA's evaluation of the lack of fire detection in the refueling area. Section 6.2.1 of this evaluation provides the NRC staff evaluation of this deviation.

Based on its review of the information submitted by TVA, the NRC staff concludes that, with the exception of items evaluated elsewhere in this evaluation, TVA's design criteria and bases related to the equipment hatches and stairwells are in accordance with the guidelines of Positions D.1.j and D.4.f of Appendix A to BTP APCSB 9.5-1 and, therefore, are acceptable.

3.1.4   Fire Doors

In FPR Part II, Section 12.10.4, TVA stated that fire door assemblies (doors, frames, and hardware) are provided for door openings required as part of fire barriers. Fire doors have been evaluated in accordance with NFPA 80-1975, "Standard for Fire Doors and Fire Windows." Fire doors are normally provided with closing mechanisms. In addition, TVA stated in FPR Part VII, Section 4.1, that some fire doors have been altered by the addition of signs and security hardware, or have been damaged and repaired onsite. Closing mechanisms and latches provided on doors are inspected to ensure proper functioning. Special purpose doors (e.g., flood, heavy equipment) installed in fire barriers have been evaluated by a fire protection engineer for acceptability.

TVA installed UL listed fire door assemblies (doors, frames, and hardware) in door openings that are required as part of fire barriers. These door assemblies are either A-labeled (3-hour), for 3-hour fire barriers or B-labeled (1½-hour), for fire barriers having a fire rating of 2 hours or less. Furthermore, TVA stated that security hardware incorporated into a fire door assembly does not adversely impact the fire rating of the assembly in accordance with NRC staff guidance in Section 3.2.3 of GL 86-10. Sliding fire doors are provided in selected locations, such as rooms protected with gaseous fire suppression systems. These sliding fire doors are closed by a fusible link or CO_2 system actuation, or both.

TVA stated that in areas protected by automatic CO_2 suppression systems, fire doors close upon the CO_2 system actuation. The thermal link on the fire doors actuates and closes prior to CO_2 fire suppression system discharge.

TVA stated that special purpose doors (e.g., air lock doors, equipment doors, and submarine-type doors) cannot be purchased as labeled fire-rated doors. FPR Part VII, Section 4.1, summarizes TVA's evaluation for the deviation of these types of fire door from the NRC staff guidance. Section 6.2.2 of this evaluation provides the NRC staff evaluation of this deviation.

Based on its review of the information submitted by TVA, the NRC staff concludes that, with the exception of items evaluated elsewhere in this evaluation, TVA's design criteria and bases related to the installation of fire doors in fire barrier assemblies are in accordance with the guidelines of Position D.1.j, of Appendix A to BTP APCSB 9.5-1 relating to the fire doors and, therefore, are acceptable.

3.1.5   Fire Dampers

Fire dampers are used to maintain the required ratings of fire-rated barriers (walls, partitions, and floors) when they are penetrated by ductwork, with the goal of preventing the propagation of fire through ducts. TVA stated that fire dampers are provided in heating, ventilation, and air conditioning (HVAC) ducts that penetrate required fire barriers. Some duct penetrations do not have fire-rated dampers and are unprotected openings. Fire dampers are provided with appropriately rated fusible links based on the ambient temperatures in the location. Fire dampers in safety-related HVAC systems may have double fusible links installed if required by a single failure analysis. Furthermore, TVA stated that ventilation openings through fire barriers required to comply with NRC regulations are protected by fire dampers having a rating equivalent to that required of the barrier. TVA stated that fire dampers have been evaluated per the requirements of NFPA 90A-1975, "Standard for the Installation of Air Conditioning and Ventilating Systems."

In areas protected by automatic CO_2 suppression systems, these dampers also close during the CO_2 system discharge. The fire dampers that provide CO_2 suppression system isolation capability are actuated by a release mechanism when the CO_2 system activates, if not actuated by a thermal link prior to CO_2 system discharge.

In FPR Part VII, Section 3.4, TVA stated that there are two instances of large fire dampers that do not meet NRC staff guidance. Section 6.2.8 of this evaluation provides the NRC staff evaluation of this deviation.

FPR Part VII, Section 3.5, summarizes TVA's evaluation for the deviation of the fire damper in the volume control tank (VCT) rooms' fire door from the NRC staff guidance. Section 6.2.10 of this evaluation provides the NRC staff evaluation of this deviation.

FPR Part VII, Section 6.2, summarizes TVA's evaluation of relaxing the surveillance frequencies for fire dampers in high radiation or contaminated areas. Section 6.3.2 of this evaluation provides the NRC staff evaluation of this deviation.

Based on its review of the information submitted by TVA, the NRC staff concludes that, with the exception of items evaluated elsewhere in this evaluation, fire dampers at WBN are installed consistent with Positions D.1.j and D.4.i of Appendix A to BTP APCSB 9.5-1, and, therefore, are acceptable.

3.1.6   Fire Barrier Penetration Seals

3.1.6.1   Electrical and Mechanical Penetration Seals

In FPR Part II, Section 12.10.6, TVA discussed seals that are installed in areas in which plant commodities, such as pipes, cable trays, conduits, etc., pass through fire rated barriers. TVA tested these seals to the time-temperature curve in American Society for Testing and Materials (ASTM) standard ASTM E119, "Standard Test Methods for Fire Tests of Building Construction and Materials," at an independent fire testing laboratory with experience in the testing of penetration seals.

The testing showed that the penetration seals could withstand the fire endurance test without the passage of flame or gases hot enough to ignite cable or fire stop material on the unexposed side for a period equal to the required fire rating. In addition, for seals required to meet other plant design bases requirements, such as radiation shielding, HVAC pressure differential, and/or flood, they were tested for such capability.

TVA stated that the penetration seal configurations at WBN have withstood a hose stream test in accordance with Institute of Electrical and Electronics Engineers (IEEE) 634-1978,

"Cable-Penetration Fire Stop Qualification Test," or ASTM E-814-83, "Standard Test Method for Fire Tests of Penetration Firestop Systems," without the hose stream causing an opening through the penetration seal that would permit a projection of water beyond the unexposed side.

TVA stated that the 1-, 2-, and 3-hour fire rated mechanical penetrations were tested in accordance with ASTM E-814-83, for the fire and "T" rating. The "T" rating acceptance criteria is limited to a temperature rise of 325 degrees Fahrenhiet (F) above ambient for cold side penetration seal surface temperatures. Service temperature and any thermal or mechanical movement of the pipe were also considered in the testing of the mechanical penetration seals.

TVA stated that 1-, 2-, or 3-hour fire rated electrical penetration seals were tested in accordance with IEEE 634-1978. Transmission of heat through the penetration seal was limited to 700 degrees F or the lowest auto-ignition temperature of cable in the penetration, whichever is lower.

Conduit penetrations that were poured in place during plant construction have internal seals. TVA stated that internal seal materials, design, and locations in walls and floor/ceiling assemblies have been evaluated as equivalent to tested configurations. For conduits with external seals (e.g., the conduits passing through a sleeve larger than the conduit), the external seal meets the same criteria as stated for electrical penetration seals.

Based on its review of the information submitted by TVA, the NRC staff concludes that the fire protection information presented in the FPR conforms to the guidelines of Positions D.1.j and D.3.d of Appendix A to BTP (APCSB) 9.5-1 and, therefore, is acceptable.

3.1.6.2   Internal Conduit Fire Barrier Penetration Seals

TVA stated that conduits that pass through fire barriers are provided with internal smoke and gas seals. TVA stated that these seals have a minimum of 3 inches of silicone foam and 1 inch of ceramic fiber damming installed at the bottom or back side of the foam seal. TVA further stated that conduits that terminate in closed junction boxes or other noncombustible sealed enclosures do not need internal smoke seals, except for conduits in the auxiliary and secondary containment envelope boundary. In addition, that an electrical cubicle, such as in a motor control center (MCC) or in a switchgear cabinet, is considered combustible and therefore would have internal conduit seals at or near the fire barrier. Conduits that are routed through the fire area and that do not terminate in the area do not have internal seals.

For lengths of conduit that extend less than 1 foot beyond the plane of a fire barrier, regardless of diameter, a fire seal is installed. For other combinations of diameters and lengths of conduit, TVA uses a graded approach for the installation of internal conduit seals, as provided in FPR Part II, Section 12.10.6. For smaller diameter conduits, a short length of conduit from the barrier is sufficient to restrict smoke or hot gases. For larger diameter conduits, longer lengths of conduit from the barriers are needed to adequately restrict the travel of smoke or hot gases.

Based on its review of the information submitted by TVA, the NRC staff concludes that TVA's criteria for the installation of internal conduit fire and smoke seals are equivalent to the guidelines of Positions D.1.j and D.3.d of Appendix A to BTP (APCSB) 9.5-1 and, therefore, are acceptable.

3.2 Safe Shutdown Capability

3.2.1 Separation of Safe Shutdown Functions

In order to ensure that one train of equipment remains free of fire damage, where components of redundant trains of systems necessary to achieve and maintain hot shutdown conditions are located within the same fire area outside the containment, TVA stated that equipment, components, cables, and associated circuits of redundant, safe shutdown systems are separated in accordance with the following separation criteria in Section III.G.2(a) through Section III.G.2(c) of Appendix R to 10 CFR 50:

(a) Separation of cables and equipment and associated non-safety circuits of redundant trains by a fire barrier having a 3-hour rating. Structural steel forming a part of or supporting such fire barriers shall be protected to provide fire resistance equivalent to that required of the barrier;

(b) Separation of cables and equipment and associated non-safety circuits of redundant trains by a horizontal distance of more than 20 feet with no intervening combustible or fire hazards. In addition, fire detectors and an automatic fire suppression system shall be installed in the fire area; or

(c) Enclosure of cable and equipment and associated non-safety circuits of one redundant train in a fire barrier having a 1-hour rating. In addition, fire detectors and an automatic fire suppression system shall be installed in the fire area;

For safe shutdown components located inside the containment building, TVA used one of the means noted above, or one of the following means to achieve separation between trains:

- fire detectors and automatic fire suppression installed in the area; or

- separation of equipment, components, and associated circuits of redundant systems by a RES

In order to conform to the fire protection and safe shutdown train separation criteria as described in Section III.G.2(b) of Appendix R to 10 CFR Part 50 listed above, TVA took credit for a safe shutdown analysis volume (AV) evaluation methodology and also took credit for enhanced automatic fire suppression consisting of pre-action sprinklers located at the ceiling level and below obstructions in the large general plant areas, and area-wide ionization smoke detection.

TVA used the AV methodology in order to sub-divide a large fire area and then subject it to a detailed safe shutdown analysis in accordance with Appendix R to 10 CFR Part 50 and ensure that one train of safe shutdown capability remains free of fire damage.

Under TVA's analysis volume methodology, an AV can consist of an entire fire area or a portion of a larger fire area. When the AV is a portion of the fire area, it can consist of multiple rooms, a single room, portions of a room (normally defined by column line locations), or any combination of the above. Each AV that involves only a portion of a room includes a 20 foot wide (minimum) "buffer zone" between it and the adjacent AV. The buffer zones are analyzed as part of the larger AV and as a separate AV. Every portion of a fire area is part of at least one AV.

In performing the safe shutdown analyses, safe shutdown components and cables are assigned to each AV containing the component. Additionally, components located in the buffer zones are assigned to an AV for the buffer zone.

TVA's safe shutdown analysis is performed assuming that all components and cables in the AV are damaged by the postulated fire. A set of safe shutdown equipment is then selected and corrective actions designated to ensure safe shutdown functions can be maintained with the selected equipment.

Some AVs in the plant use electrical raceway fire barrier system (ERFBS) for redundant trains located within a single AV. The ERFBS extends to the boundary of the AV to assure separation between redundant trains within the AV. For large AVs, this may not be a barrier; rather it may be the column line or other indicator of the edge of the AV.

In order to provide reasonable assurance that WBN satisfied the technical requirements in Section III.G, "Fire Protection of Safe Shutdown Capability," of Appendix R to 10 CFR Part 50, TVA identified and used the following types of analysis volumes, as described with figures in FPR Part III, Section 10.3:

- *Fire Area* - The fire area is separated from other adjacent areas by rated barriers (walls, floors, and ceilings) that are sufficient to withstand the hazards associated with the area and, as necessary, to protect equipment in the area from a fire outside the area.

- *Single Room within a Fire Area* - A room may be separated from other adjacent rooms in a fire area by regulatory fire barriers (walls, floors, and ceilings) that have a 1-hour or greater fire rating.

- *Combination of Rooms within a Fire Area* - The combination of rooms in the AV are separated from other AVs within the same fire area by regulatory fire barriers that are rated for at least 1-hour

- *Sections of Large General Areas* - AVs consisting of sections of large general areas are separated from each other by "buffer zones" that are wider than 20 feet. In large general areas where buffer zones are used that include intervening combustibles, enhanced automatic suppression and detection systems are installed in the large general area. Where AVs are separated from other AVs by buffer zones, a fire in one of the AVs would not be expected to pass through the buffer zone and affect equipment in the AV on the other side of the buffer zone. TVA uses combinations of overlapping AVs in their analysis.

- *Sections of Large Rooms* - For AVs that consist of large room sections separated by an overlap region that is greater than 20 feet, the overlap region is considered to be part of both AVs. If the overlap region contains intervening combustibles, enhanced automatic suppression and detection systems are installed in the large room.

For large general areas and large rooms that have either buffer zones or overlap regions, refer to Section 6.1.4 of this evaluation for additional information regarding fire protection in those regions.

Based on its review of the submitted information, the NRC staff concludes that TVA's criteria for providing fire protection for safe shutdown functions provides an equivalent level of fire safety to Section III.G. of Appendix R to 10 CFR Part 50 and is, therefore, acceptable.

3.2.2   Safe Shutdown - General Plant Areas

TVA's methodology for assessing compliance with the separation/protection requirements of Section III.G of Appendix R to 10 CFR Part 50 consisted of:

(a) Determining the functions required to achieve and maintain safe shutdown

(b) Producing shutdown logic diagrams that define minimum sets of systems capable of accomplishing each shutdown function

Each plant system or subsystem function relied on to accomplish the above safe shutdown functions is identified. A separate designator is assigned to each plant system or subsystem function to ensure consistency between analysis documents and calculations. Each designator is identified as a safe shutdown "Key." The safe shutdown logic diagram (FPR Figure III-5) depicts the safe shutdown system and/or system function, associated Key number, and logical relationships between systems and Keys used to demonstrate compliance with the criteria in Appendix R to 10 CFR Part 50.

(c) Grouping specific plant locations into fire areas

(d) Identifying for each area, one or more paths through the shutdown logic diagrams that satisfy each required shutdown function

(e) Developing functional criteria that defined the required equipment for the shutdown paths

(f) Identifying power and control cables for shutdown-related equipment and associated circuits that are not isolated from shutdown cabling

For each safe shutdown key, cable block diagrams were developed for each safe shutdown component to identify cables required to ensure that the component can perform its safe shutdown function. Raceways that contain these required cables were then identified, and their locations documented. An interaction is defined as a place in the plant where redundant safe shutdown paths are not separated in accordance with the requirements in Section III.G.2 of Appendix R to 10 CFR Part 50. Whenever an interaction was identified, it was documented and evaluated for its impact on safe shutdown capability. An appropriate resolution was then determined and documented.

(g) Resolutions may consist of modifications, use of alternate equipment, OMAs, fire barrier or radiant energy shield installation, post-fire repairs, engineering evaluations prepared in accordance with the guidance in GL 86-10, or deviation requests

Based on its review of the information submitted by TVA, the NRC staff concludes that TVA's methodology for assessing compliance with the separation/protection requirements in Section III.G of Appendix R to 10 CFR Part 50, is acceptable.

3.2.3   Safe Shutdown Analysis

TVA stated that its safe shutdown analysis demonstrated that sufficient redundancy exists for systems needed for hot and cold shutdown. The safe shutdown analysis included components, cabling, and support equipment needed to achieve hot and cold shutdown.

TVA stated that for hot shutdown, at least one train of the following safe shutdown systems would be available: (1) auxiliary feedwater (AFW) system, (2) steam generator (SG) power-operated relief valves (PORVs), (3) reactor coolant system (RCS), and (4) chemical and volume control system. For cold shutdown, at least one train of the residual heat removal (RHR) system would be available. TVA stated that the RHR system provides the capability to achieve cold shutdown within 72 hours after a fire, and would be used for long-term decay heat removal. The availability of these systems includes the components, cabling, and support equipment necessary to achieve cold shutdown. Support equipment includes the diesel generators and associated electrical distribution system, the essential raw cooling water (ERCW) system, the component cooling water system (CCS), and the necessary ventilation systems.

TVA stated that an electrical separation study was performed to ensure that at least one train of such equipment is available in the event of a fire in areas that might affect these components. Safe shutdown equipment and cabling were identified and traced through each fire area from the component to the power source. Associated circuits whose fire-induced spurious operation could affect safe shutdown were identified by a system review to determine those components whose maloperation could affect safe shutdown capability. The potential for MSO was also analyzed. Further discussion of the MSO is presented below in Section 3.9, "Assessment of Multiple Spurious Operations."

TVA stated that alternative shutdown measures are required only for fires in the control building. If a fire disables the MCR or requires the evacuation of the MCR, the ACR, which is located in a separate fire area in the auxiliary building, would be available to achieve and maintain the plant in hot standby and subsequent cold-shutdown conditions. The control functions and indications provided at the ACR panel are electrically isolated or otherwise separate and independent from the MCR. Further discussion of the alternative shutdown capability is presented below in Section 3.3, "Alternative Shutdown."

Based on its review of the information submitted by TVA, the NRC staff concludes that the systems identified by TVA for achieving and maintaining safe shutdown in the event of a fire as described in Section III.G of Appendix R to 10 CFR Part 50 are acceptable.

3.2.4   Systems Required for Safe Shutdown

TVA stated that shutdown of the reactor and reactivity control is initially performed by control rod insertion. Long term reactivity control is provided by adding borated water from the refueling water storage tank (RWST). RCS inventory is maintained by varying charging and letdown flow through the RCS makeup and letdown paths. Decay heat removal during hot shutdown is accomplished by establishing secondary-side pressure control and supplying water to two of the four SGs from one of the redundant motor- or turbine-driven AFW pumps. Long-term heat removal to establish and maintain cold-shutdown conditions is provided by the RHR system.

TVA stated that primary system pressure is controlled by the pressurizer heaters (if available) or by varying pressurizer level in combination with control of SG pressure and RCS temperature using SG PORVs.

Based on its review of the information submitted by TVA, the NRC staff concludes that the systems selected by TVA are capable of satisfying the post-FSSD requirements in Sections III.G and III.L of Appendix R to 10 CFR Part 50, and therefore, are acceptable.

3.3   Alternative Shutdown

3.3.1   Areas in Which Alternative Shutdown Is Required

TVA's analysis identified that alternative shutdown capability is required for control building fires that also require shutdown from outside of the MCR. For these fires, cold shutdown must be achieved within 72 hours. TVA also indicated that it evaluates the alternative shutdown capability in accordance with Sections III.G.3 and III.L of Appendix R.

3.3.2   Alternative Shutdown System

The alternative shutdown system uses existing plant systems and equipment identified in Section 3.2 above, and an ACR complex. TVA stated that the analysis indicates that for control building fires, no repairs are required to implement the alternative shutdown capability.

A loss of offsite power is required to be postulated for those locations that require alternative shutdown. TVA stated that the systems used during alternative shutdown are can be powered by both onsite and offsite power.

The ACR complex is physically independent of the control building. Where required, electrical isolation of controls and indications provided for the ACR is achieved through the actuation of isolation/transfer switches. The ACR complex is divided into five independent rooms consisting of a Train A and Train B transfer switch room for each unit and the ACR. The ACR serves as the central control point during alternative shutdown from outside the MCR, and provides control and monitoring capability for redundant trains (Trains A and B) of equipment required to achieve safe shutdown.

TVA also analyzed the potential for MSOs. Section 3.9 of this safety evaluation further discusses MSOs.

3.3.3   Alternative Shutdown Conclusion

Based on its review of the information submitted by TVA, the NRC staff concludes that the alternative shutdown system is consistent with Sections III.G.3 and III.L of Appendix R to 10 CFR Part 50, and therefore, is acceptable.

3.4   Alternative Shutdown Performance Goals

TVA stated that the alternative shutdown system described in Sections 3.4.1 through 3.4.5 was designed to enable the achievement of alternative shutdown performance goals outlined in Section III.L of Appendix R to 10 CFR Part 50.

3.4.1   Reactivity Control

Initial reactivity control is provided by the control rods, which are inserted by the reactor protection system. Additional shutdown margin is provided by injecting borated water from the RWST into the RCS via the charging pumps. Source range monitoring instrumentation is available in the ACR to monitor reactivity and to ensure adequate shutdown margin.

3.4.2   Reactor Coolant Inventory

Control of the RCS inventory requires maintaining the reactor coolant pump (RCP) seal integrity and RCS pressure boundary integrity and providing RCS makeup and letdown.

RCP seal cooling is required to maintain seal integrity and to prevent an uncontrolled loss of reactor coolant inventory. Diverting a portion of the charging flow to the RCP seals achieves RCP seal cooling. Isolating the normal and excess letdown lines, in turn, isolates the RCS pressure boundary. To prevent depressurization of the RCS, the plant ensures that the solenoid valves in the reactor vessel head vent system remain closed.

RCS inventory is controlled by varying charging and letdown flow through RCS makeup and letdown paths. One of the redundant centrifugal charging pumps is required to provide makeup inventory to the RCS. The VCT is required to provide a short-term supply of water for makeup of RCS inventory and RCP seal cooling. A suction path from the RWST is required to provide a long-term source of borated water for RCS makeup. If necessary, inventory may be removed from the RCS by way of the pressurizer PORVs, discharging to the pressurizer relief tank (PRT), or discharging through the RCS head vent valves.

Reactor coolant makeup is usually available immediately following reactor trip from the charging system, except in a few fire locations where it is available within 75 minutes following reactor trip. TVA stated that an analysis was performed which demonstrates that makeup due to RCS leakage is not required for 75 minutes. TVA stated that for these scenarios, maintaining the RCS integrity is necessary to achieve adequate inventory control. The inadvertent opening of boundary isolation valves, such as the reactor head vent valves and RHR suction isolation valves, has been precluded, and adequate RCP seal integrity is maintained to assure safe shutdown.

3.4.3   Decay Heat Removal

RCS temperature from power operation to hot-shutdown conditions is controlled by the rate of heat removal from the reactor coolant to the secondary-side coolant and from hot shutdown to cold shutdown via direct heat transfer by the RHR system to the ultimate heat sink. During RCS cooldown to RHR entry conditions, heat will be removed from the reactor and transferred to the SGs via natural circulation. The removal of decay heat for cooldown from reactor trip to hot standby conditions requires one AFW pump supplying water to two of the four SGs. The required makeup water supply can come from either the condensate storage tank (CST) or from ERCW.

The CST is normally aligned to the suction of the AFW pumps. WBN is supplied with two motor-driven AFW pumps per unit with only one per unit required for safe shutdown. The turbine-driven AFW pump (one per unit) is designed to deliver a sufficient flow to all four SGs and maintain SG water levels at the lower limit of the wide range level indicator.

The RHR system is required to provide the long-term heat removal capability necessary to establish and maintain cold-shutdown conditions. The establishment of RHR cooling requires one RHR pump, a heat exchanger, and the associated flowpath to provide RCS coolant flow to the primary side of the RHR heat exchanger; one CCS pump and its associated flowpath to provide cooling to the secondary side of the RHR heat exchanger; and one ERCW pump and its associated flowpath to supply cooling water to the CCS heat exchanger. If the diesel generators are required to supply required power, an additional ERCW pump would be required for cooling purposes.

TVA's post-fire shutdown analysis states that the pressurizer heaters are the preferred method of controlling RCS pressure, and will be used if available. If the pressurizer heaters are not available, RCS pressure can be controlled by controlling pressurizer level using the charging system.

3.4.4   Process Monitoring

Direct indication of process variables including reactor coolant hot-leg temperature (T-hot), reactor coolant pressure, pressurizer level, SG level and pressure, source range flux, charging header pressure and flow, VCT level indication, and decay heat removal system flow are provided in the ACR.

TVA requested a deviation to Appendix R requirements for instrumentation necessary to achieve alternative shutdown. Specifically, contrary to Appendix R requirements, TVA has not provided wide-range SG level, tank level indication for the condensate and RWSTs, and RCS cold-leg temperature (T-cold). Section 6.1.1 of this evaluation provides the NRC staff evaluation of this deviation.

3.4.5   Support Functions

The FPR and the associated shutdown logic diagram (FPR Figure III.5) identify the emergency power distribution system, offsite power system, ERCW system, CCS, HVAC to areas containing essential FSSD equipment, and control room chillers as required support functions.

TVA stated that this essential HVAC is provided for the control, auxiliary, diesel generator, and reactor buildings. Portions of the systems in each building that service safe shutdown equipment required for compliance with Appendix R have been analyzed to ensure that at least one path of the required systems will be available for an Appendix R fire. These systems include the primary safety-related portions of the control building, the auxiliary building HVAC system for the 480V transformer rooms and for the general floor area on the 713.0 foot elevation, the turbine-driven AFW pump room, the diesel generator HVAC systems including the diesel generators, associated batteries and electrical boards and the containment air cooling systems. All other areas of the plant which contain equipment required for safe shutdown per Appendix R have been evaluated and determined that acceptable temperatures will be maintained for the required equipment to perform its intended function if HVAC is lost.

3.4.6   Alternative Shutdown Performance Goals Conclusion

Based on its review of the information submitted by TVA, the NRC staff concludes, with the exception of items evaluated elsewhere in this evaluation that TVA's treatment of alternative shutdown performance goals is consistent with Section III.L of Appendix R to 10 CFR Part 50, and therefore, is acceptable.

3.5   Operator Manual Actions

TVA's post-FSSD analysis, and associated cable interaction studies, identified some fire areas where operator actions to take manual control of equipment may be required to compensate for fire-induced equipment failures. TVA classified OMAs into two general categories: (1) manual actions for safe shutdown success path SSCs and (2) manual actions for SSCs important to safe shutdown. Repairs for cold shutdown are also included, but are not considered OMAs.

TVA referenced Revision 2 to RG 1.189 for the discussion of safe shutdown success path SSCs and SSCs important to safe shutdown.

3.5.1   OMAs for Safe Shutdown Success Path SSCs

In FPR Part V, Section 2.0, TVA stated that OMAs for SSCs in the safe shutdown success path require prior NRC approval. The position that OMAs for SSCs in the safe shutdown success path require prior NRC approval is consistent with the guidance in Revision 2 to RG 1.189. WBN Unit 1 OMAs for success path SSCs were approved in NRC SSER 18, prior to operation of Unit 1. The TVA evaluations of WBN Unit 2 OMAs for success path SSCs are included in FPR Part VII, Section 8. Section 6.1.9 of this evaluation provides the NRC staff evaluation of these OMAs.

TVA stated that future safe shutdown success path SSCs OMAs, or such OMAs for WBN Unit 1 implemented since the SSER 18, will be submitted to the NRC for approval, consistent with the language in the FPR.

3.5.2   OMAs for SSCs That Are Important to Safe Shutdown

In FPR Part V, Section 2.0, TVA stated that OMAs for SSCs that are important to safe shutdown do not require prior NRC review and approval. The position that OMAs for SSCs that are important to safe shutdown do not require prior NRC approval is consistent with the guidance in Revision 2 to RG 1.189. Area-specific evaluations for any area where WBN Unit 2 OMAs involving important to safe shutdown equipment that are need to be performed in the area of fire origin are evaluated in Section 3.5.6 below.

TVA discussed the feasibility and reliability analysis criteria for evaluating OMAs. In FPR Part V, Section 2.1, TVA stated that these criteria are based on NUREG-1852.

For all important to safe shutdown SSC manual actions, TVA considered defense-in-depth features, such as fire prevention (transient combustible and hot work controls), fire detection and suppression, and area separation. For any area crediting an OMA with less than 2 hours of required time, and that also lacks robust defense-in-depth fire protection features; additional time margin is included in addition to the nominal acceptance criteria.

TVA considers the following factors in its evaluation of these OMAs: (1) time, (2) environmental factors (smoke, lighting, noise, etc.), (3) necessary equipment, (4) procedures, and (5) staffing. Each of the factors included acceptance criteria. For example, all OMAs have an allowable time of 10 minutes or greater with 100 percent margin. Factors that could cause delays in the performance of the OMA have also been considered. Factors such as lighting and communications are supported by plant calculations.

TVA evaluated the access routes necessary to perform the OMAs. Because some areas of the plant are separated into separate AVs, it is possible that OMAs may occur in a portion of a fire area that is remote to the portion where fire damage could affect safe shutdown equipment. In this event, additional access routes have been evaluated. TVA walked down these alternative access routes and determined that they are viable.

TVA used current NRC guidance to develop acceptance criteria for OMAs for SSCs that are important to safe shutdown. TVA incorporated a review of defense-in-depth, feasibility, and reliability in their analysis.

Based on its review of the submitted information, the NRC staff concludes that OMAs for SSCs important to safe shutdown include consideration of defense-in-depth, feasibility, and reliability, and therefore, this approach is acceptable.

3.5.3   OMAs Required Prior to Control Room Evacuation

The only operator action normally credited prior to MCR evacuation is a reactor trip (scram). However, the FPR credits two actions prior to MCR evacuation: (1) closing the two pressurizer PORV block valves and (2) tripping the RCPs. TVA stated that closing the PORV block valves is needed to prevent loss of RCS pressure/inventory due to possible spurious PORV opening prior to transferring plant control to the ACR. Also, an immediate trip of the RCPs is necessary to prevent overcooling caused by a spurious actuation of pressurizer spray valves, whose circuits are not isolated from the control building.

TVA evaluated the feasibility and adequacy of the proposed approach for performing three distinct actions prior to MCR evacuation: (1) scram, (2) PORV block valve closing, and (3) RCP trip. The evaluation assumed that a fire in the MCR would be characterized by slow growth and be detected in its early stages by control room operators or installed smoke detection systems. Fires in other areas of the control building may require MCR evacuation; such as in the cable spreading room or auxiliary instrument room, etc. The control building areas other than the MCR have installed detection and automatic suppression systems, or have a deviation documented in FPR Part VII, Section 2.3. TVA stated that areas of the control building that don't have automatic suppression typically have low combustible loading.

In NRC question RAI FPR V-16, the NRC staff expressed a concern that a fire in portions of the control building that lack fire detection and automatic suppression could impact equipment important to safe shutdown. In TVA's letter dated August 5, 2011 (ADAMS Accession No. ML11224A052), TVA stated that the PORV block valve controls are only routed through control building areas that have detection and automatic suppression. Other circuits routed through the control building are either routed through areas with fire detection and automatic suppression, or areas with detection and limited combustibles and ignition sources.

Based on its review of the submitted information, the NRC staff concludes that TVA considered the credible fire scenarios, and the installed defense-in-depth features to determine that these three control room actions are feasible, therefore, the performance of these three distinct actions is acceptable.

3.5.4   Safe Shutdown Procedures and Manpower

TVA developed a fire response procedure, Abnormal Operating Instruction (AOI)-30.1, "Plant Fires," which describes operator response and mitigating actions for plant fires. TVA also developed room-specific procedures as part of AOI-30.2, "Fire Safe Shutdown," for rooms where OMAs may be required to mitigate damage to plant safe shutdown equipment. AOI-30.2 is supported by controlled plant calculations. The procedures include operator-by-operator actions for a fire in any room of the plant that would require OMAs to shutdown the plant.

TVA has walked down the OMAs for both WBN Unit 1 and WBN Unit 2. OMAs needed after 2 hours into the fire were not walked down, since 2 hours corresponds to the time frame for additional personnel to be called to the plant in response to an event. TVA postulates that significant plant fires are interior to the plant; therefore, operators who are called back are not expected to have difficulty getting to the plant.

TVA stated that the start of the time "clock" for the performance of OMAs is the tripping of the plant. Prior to tripping the reactor, the plant is considered to be in a stable operating condition. Once the trip is initiated, the clock starts and preventive OMAs are performed to prevent spurious equipment operation and to ensure safe shutdown can be accomplished.

Most OMAs are preventive, however, some reactive OMAs must be taken upon fire damage to SSCs rather than reactor trip. TVA stated that for these reactive type actions, the normal plant operating procedures provide an appropriate reactive response to fire damage. TVA analyzed the available FSSD equipment on an area by area basis to assure that sufficient safe shutdown equipment is free of fire damage.

Based on its review of the submitted information, the NRC staff concludes that TVA's safe shutdown procedure structure, including both preventive and reactive OMAs, has been evaluated to ensure safe shutdown capability, and therefore, is acceptable.

3.5.5   Repairs

TVA stated that repair activities (e.g., lifting/cutting leads, installing jumpers, and fuse replacement) are not required to achieve and maintain hot standby conditions. TVA identified the following three generic repairs to be performed to achieve cold shutdown:

- Loading Two ERCW Pumps on 6.9 kV Board 1-BD-211-A-A,
- RHR Room Cooler Repair, and
- RHR/RCS High-Low Pressure Boundary Valve Repair.

Cold-shutdown repair activities include the installation of electrical jumpers, and the installation of replacement cables and components if needed due to fire damage. TVA has identified the specific activities to be performed and has developed repair procedures to implement this capability. Additionally, materials necessary to accomplish the repairs are available on site.

Based on its review of the submitted information the NRC staff concludes that the repair activities developed by TVA to achieve cold shutdown conditions are consistent with the requirements in Appendix R to 10 CFR Part 50 and therefore, are acceptable.

3.5.6   Unit 2 OMAs Involving Fire Area Re-Entry

TVA examined Unit 2 OMAs that involve re-entry into plant fire areas. This section discusses actions involving important to safe shutdown equipment, whereas Section 6.1.9 of this evaluation addresses OMAs involving equipment required for safe shutdown.

TVA has indicated that all WBN Unit 2 rooms that involve re-entry to perform OMAs for important to safe shutdown equipment are equipped with fire detection and automatic suppression. In addition, TVA stated that all such OMAs have approximately 60 minutes or more of time margin for the licensee staff to extinguish the fire and to operate the equipment within the room. TVA has determined that the equipment within the room of fire origin is unlikely to be damaged such that the equipment could not be operated following a postulated fire in that room. TVA has performed feasibility and reliability evaluations of the OMAs.

Based on its review of the submitted information, the NRC staff concludes that there is sufficient defense-in-depth available, detection and automatic suppression is installed, and that the manual action provides sufficient margin to assure safe shutdown capability. Therefore the NRC staff concludes that such re-entry into rooms to perform OMAs involving important to safe shutdown components is acceptable.

3.6   Associated Circuits

TVA examined the potential impact of fire damage on associated circuits of concern. TVA has categorized associated circuits as follows:

- Type I – common power source,
- Type II – spurious actuation, and
- Type III – common enclosure.

TVA stated that it identified these associated circuits of concern in accordance with GL 81-12, the NRC staff's clarification to GL 81-12, and GL 86-10.

3.6.1   Circuits Associated by Common Power Source

TVA stated that, for circuits associated by a common power source, it has identified all circuits supplied from a power source (i.e., switchgear, MCCs, and load centers) that also powers a circuit of equipment required for post-FSSD. For the identified circuits, it verified the coordination of electrical protection devices (e.g., fuses, circuit breakers, or relays) to ensure that a fire-induced fault on a branch circuit of a required supply will be cleared by at least one branch circuit protective device before the fault current can propagate to cause a trip of any feeder breaker upstream of the required supply.

In its letter dated August 5, 2011 (ADAMS Accession No. ML11227A257), in response to the RAI FPR III-17, TVA stated that a list of the design change packages has been issued to ensure that the WBN Unit 2 circuits are adequately protected with fuses/breakers to address common power supply and common enclosure associated circuits of concern. Additionally, the plant will implement these design change packages in accordance with their associated system turnover schedule and implement them before the associated system being declared operable to support WBN Unit 2 fuel load or startup.

TVA evaluated circuits associated by a common power source for multiple high impedance faults (MHIFs). TVA stated that MHIFs are evaluated in accordance with the base case conditions in Appendix B.1 to Nuclear Energy Institute (NEI) 00-01, "Guidance for Post Fire Safe Shutdown Circuit Analysis," Revision 2, issued May 2009, as endorsed by Section 5.5.2 of Revision 2 to RG 1.189. The base case set of conditions, if met, provides reasonable assurance that MHIFs will not occur. The FPR, Part III, Section 7.4, analysis provided the NEI 00-01 base case conditions with the corresponding WBN compliance method for each base case condition. The

FPR stated: "WBN meets the NEI 00-01, Appendix B.1 base case criteria which establish applicability of the base case to individual plant designs." WBN did not take any exceptions to the base cases. In a letter dated June 27, 2012 (ADAMS Accession No. ML12181A531), TVA provided a list of supporting calculations for the FPR, Part III, Section 7.4, MHIF analysis.

Based on its review of the information submitted by TVA, the NRC staff concludes that TVA's method of evaluating circuits associated by a common power source is consistent with the NRC guidance in the GLs identified in Section 3.6 above, and in Appendix B.1 to NEI-00-01, as endorsed by RG 1.189; and therefore, is acceptable.

3.6.2   Circuits Associated by Spurious Operation

TVA stated that cables that are not part of safe shutdown circuits may be damaged by the effects of postulated fires. This cable damage may consequently prevent the correct operation of safe shutdown components, or result in the maloperation of equipment which would directly prevent the proper performance of the safe shutdown systems. The effects of spurious operations may be conceptually divided into two subclasses as follows:

(1) maloperation of safe shutdown equipment due to control circuit electrical interlocks between safe shutdown circuits and other circuits (e.g., the numerous safe shutdown equipment automatic operation interlocks from process control and instrument circuits)

(2) maloperation of equipment that is not defined as part of the safe shutdown systems, but that could prevent the accomplishment of a safe shutdown function (e.g., inadvertent depressurization of the RCS or the main steam system by spurious opening of boundary valves)

TVA performed an evaluation of Appendix R to 10 CFR Part 50 events to ensure that any failure of associated circuits of concern by spurious operation will not prevent safe shutdown. Credible electrical faults considered in the analysis included open circuit, short circuit (conductor-to-conductor), short to ground, and cable-to-cable (hot-short) including 3-phase hot-shorts for high/low pressure interface valves. The analysis also considered that the normally ungrounded 125 VDC power distribution system may become grounded due to fire damage.

TVA indicated that these Type II associated circuits of concern outside of containment are analyzed in accordance with the criteria in Sections III.G.2.a, III.G.2.b, and III.G.2.c of Appendix R to 10 CFR Part 50 as required circuits. Inside containment, the Type II associated circuits of concern are analyzed in accordance with the criteria in Sections III.G.2.d, III.G.2.e, and III.G.2.f of Appendix R to 10 CFR Part 50 as required circuits.

Based on its review of the information submitted by TVA, the NRC staff concludes that TVA's approach to analyze circuits associated by spurious operation, in accordance with the sections of Appendix R to 10 CFR Part 50 listed above, is acceptable.

3.6.3   Circuits Associated by Common Enclosure

To address the common enclosure associated circuit concern, TVA evaluated all circuits that may share a common enclosure (e.g., cable tray, conduit, panel or junction box) with a circuit required by Appendix R to 10 CFR Part 50. On the basis of its evaluation, TVA concluded that the electrical protective equipment provided will ensure that electrical faults and overloads will not result in any more cable degradation than would be expected when operating conditions are below the setpoint of the electrical protective device.

TVA stated that the plant addressed associated circuits by common enclosure by ensuring that all required existing (prior to 1995) circuits in buildings with safe shutdown components are electrically protected with a fuse or breaker that will actuate prior to the jacket of existing faulted cables from reaching their auto-ignition temperature. Additionally, for new circuits, associated circuit electrical fault protection is provided to ensure that the fuse or breaker will operate prior to the temperature of the insulation reaching its insulation damage temperature.

In its letter dated August 5, 2011 (ADAMS Accession No. ML11227A257), in response to the RAI FPR III-17, TVA stated that a list of the design change packages has been issued to ensure that the WBN Unit 2 circuits are adequately protected with fuses/breakers to address common power supply and common enclosure associated circuits of concern. Additionally, these design change packages will be implemented in accordance with their associated system turnover schedule and will be implemented prior to the associated system being declared operable to support WBN Unit 2 fuel load or startup.

Based on its review of the information submitted by TVA, the NRC staff concludes that TVA's methodology for assessing circuits associated by common enclosure is consistent with the NRC guidance in the GLs identified in Section 3.6 of this safety evaluation, and, therefore, is acceptable.

3.7   Current Transformer Secondaries

Section III.G.2 of Appendix R to 10 CFR Part 50 requires that fire induced open circuits be analyzed where they could prevent operation or cause maloperation of components required for post-FSSD.

If a fire at a remote location causes the secondary circuit of a current transformer (CT) to open, the event could generate ionized gases or additional fires, or both, in other locations and could propagate fire to additional fire areas.

TVA evaluated the fire hazards due to a fire-induced open circuit in the secondary circuits of CTs installed in high energy panels (i.e., 6.9 kV switchgear) of the required power systems. An evaluation of three types of CT circuits used in the auxiliary power system has been done: (1) ground fault, (2) differential relaying, and (3) protective relaying.

The CT circuits are contained in their respective panels for the Appendix R to 10 CFR Part 50 required and nonrequired 480 V switchgear and the 6.9 kV switchgear. Therefore, the fire would have to be localized in the switchgear assembly for the CT secondary circuit to be opened by a fire. This would prevent the CT circuits from causing fire propagation to other fire areas.

The 6.9 kV CT circuit that is connected to protective relaying and a current transducer is also contained within the switchgear panel. The output of the current transducer is connected to a remote indicator, and the current transducer is an electrical isolator. Additionally, the output-to-input of the current transducer has been tested for 1500V AC differential. Electrical isolation also exists for the Watt & VAR transducers used on the 6.9 kV switchgear at WBN.

The board differential relaying circuits are totally internal to the switchgear panels, except for the following three exceptions:

(1) The circuits between the 6.9 kV switchgear emergency supply feeders and the diesel generators are included in the interaction analysis as required circuits. The protective relays are designed to operate and clear these circuits in case of fire damage.

(2) The common station service transformers transformer differential relaying circuits are also included in the interaction analysis as required circuits. The current imbalance created by an open CT circuit causes the protective differential relay to open the supply circuit breaker, which removes primary power to the CT, clearing the circuit, within the time required for protective relay and breaker operation.

(3) The circuits between the 6.9 kV start and Unit Boards, are not required circuits. Similar to Item (2) above, current imbalance in the protective differential relay of the non-required circuits would open the supply circuit breaker.

Based on its review of the information submitted by TVA, the NRC staff concludes that TVA's approach to evaluating the fire hazards due to fire-induced open circuits in the secondary of CTs installed in high energy panels is in accordance with Section III.G.2 of Appendix R to 10 CFR Part 50, and therefore, is acceptable.

3.8   High/Low-Pressure Interfaces

TVA stated that GL 81-12, GL 86-10, and Information Notice (IN) 87-50, "Potential LOCA at High- and Low-Pressure Interfaces from Fire Damage," dated October 9, 1987, describe special considerations for high/low pressure interfaces that are necessary to meet the requirements in Appendix R to 10 CFR Part 50.

In accordance with GL 81-12, the following information is necessary to ensure that high/low pressure boundary interfaces are adequately protected for the effects of a single fire:

(1) Identify each high/low pressure interface that uses redundant electrically controlled devices (such as two series motor operated valves) to isolate or preclude rupture of any primary coolant boundary.

(2) Identify the essential cabling for each device.

(3) Identify each location where the identified cables are separated by a barrier having less than a 3-hour fire rating.

(4) For the areas identified in (3) above (if any), provide the bases and justification.

Based on the above, TVA performed a review of the systems credited for safe shutdown to identify potential high/low pressure interfaces. These interfaces were evaluated to identify valves that, if spuriously opened, would expose low pressure piping to high pressure resulting in failure of the low pressure system.

The control system for RHR valves has been designed to prohibit opening unless the reactor coolant pressure is low enough to prevent RHR piping failure. However, if these valves opened spuriously, exposure of RHR piping to high pressure may cause failure of the RHR system piping and render the system inoperable. Therefore, the RHR/RCS isolation valves (1/2-FCV-74-1, -2, -8, and -9) are considered high/low pressure interface valves.

Excess letdown is not required for safe shutdown. However, the spurious opening of these valves could expose downstream piping to excess pressure that may cause failure resulting in the rupture of the primary coolant boundary. Therefore, the excess letdown isolation valves (1/2-FCV-62-55, and -56) are considered high/low pressure interface valves.

Normal letdown is not required for safe shutdown. However, spurious opening of these valves may cause failure to maintain RCS inventory control. Therefore, the normal letdown isolation valves (1/2-FCV-62-69A and -70A) are considered high/low pressure interface valves.

The safety injection system (SIS)/RHR interface valve with the RCS is located in piping that connects the SIS with the RHR system at a point between the RCS/RHR isolation valves. The SIS is not required for safe shutdown. However, the spurious opening of valve 1/2-FCV-63-186 along with either 1/2-FCV-74-1-A or -9-B could expose the SIS piping to damaging pressure. Therefore, this valve is considered a high/low pressure interface.

The pressurizer PORV and reactor head vent isolation valves are designed to function at high RCS operating pressure. They provide the following two safe shutdown functions: (1) to initially remain closed for RCS inventory control purposes and (2) to provide a means of depressurizing the RCS to the point that the RHR system can be initiated to bring the plant to a cold shutdown condition. Discharge from the RCS through these valves is directed to the inlet of the PRT. The inlet lines are sized to accommodate vent/relief discharge flow without piping or component failure. Continuous letdown to the PRT may eventually cause spillage of excess coolant to containment through the PRT rupture disks. Therefore, the pressurizer PORVs and block valve combinations, and reactor head vent isolation valves, are required for RCS inventory control (and RCS letdown) and are considered high/low interface valves.

To prevent fire-initiated cable faults from causing a spurious operation of the RHR isolation valves, all four of the motor operated valves in the RHR suction line will be kept closed (pre-fire condition) with the corresponding MCC breaker in the open position. The return lines are isolated by two series check valves in each line and a common motor-operated valve.

In its letter dated May 30, 2012 (ADAMS Accession No. ML12153A374), TVA stated that procedural controls for isolation of all potentially spurious RCS letdown paths, including pressurizer PORVs and reactor head vents, provide assurance (through the use of MCR actions for WBN Unit 2 and MCR actions and an OMA for WBN Unit 1), that isolation of normal and excess letdown paths will be achieved.

Based on its review of the information submitted by TVA, the NRC staff concludes that TVA's approach for high/low pressure interfaces meets the requirements in Appendix R to 10 CFR Part 50 and follows the guidance in GL 81-12 and GL 86-10, and IN 87-50, and therefore, is acceptable.

3.9   Assessment of Multiple Spurious Operations

In FPR Part III, Section 11.0 "Multiple Spurious Operation (MSO) Evaluation," TVA stated that Revision 2 to RG 1.189 formalized the requirements for addressing multiple fire induced circuit failures, or MSOs and multiple concurrent hot shorts. TVA further stated that this process was followed to address fire-induced spurious failures for WBN Unit 1. In a letter dated August 20, 2010 (ADAMS Accession No. ML102360283), TVA stated that the MSO scenarios requiring resolution for WBN Unit 1 would be implemented under the timing requirements prescribed by the NRC in Enforcement Guidance Memorandum 09-002, "Enforcement Discretion for Fire Induced Circuit Faults."

In TVA MSO Evaluation R-1976-20-001, "Watts Bar Nuclear Plant Unit 2 Multiple Spurious Operation Evaluation, Revision 1" (ADAMS Accession No. ML103160419), TVA stated that multiple fire-induced spurious failures were evaluated at WBN Unit 2 as described in Revision 2 to RG 1.189. TVA further stated that, based on the results of the MSO expert panel conducted at the plant for WBN Unit 1, various scenarios were identified and were reviewed for WBN Unit 2. Appendix B, "Unit 2 Resolutions," and Appendix C, "Unit 1/Common Resolutions," of the above report provided resolutions for specific unresolved MSO scenarios that affect WBN Unit 2.

In a letter dated February 7, 2013 (ADAMS Accession No. ML13044A114), TVA stated that the above MSO scenarios requiring resolution for WBN Unit 1 have been resolved and incorporated into Appendix B and Appendix C to Revision 2 to MSO Evaluation R-1976-20-001.

In Section 4 of the MSO Evaluation (Revisions 1 and 2), TVA stated that MSO scenarios selected for Sequoyah Nuclear Plant, Units 1 and 2, and for WBN Unit 1 were evaluated to determine if the scenarios were applicable to WBN Unit 2 and how Unit 2 complied with each scenario. Sequoyah Nuclear Plant, Units 1 and 2, and WBN Units 1 and 2, have similar physical and systems designs. All four units are Westinghouse four-loop pressurized water reactors with wet ice condenser containments and would be expected to have similar MSO scenarios. Additionally, the Sequoyah Nuclear Plant, Units 1 and 2, MSO scenarios were analyzed from a dual unit perspective.

Based on its review of the information submitted by TVA, the NRC staff concludes that by evaluating multiple fire-induced spurious failures in accordance with the guidance in Revision 2 to RG 1.189 and by using MSO scenarios from Sequoyah Nuclear Plant, Units 1 and 2, and WBN Unit 1 when addressing WBN Unit 2 and dual-unit scenarios, TVA's approach is an acceptable means for addressing MSO failures.

3.10   Smoke Control and Ventilation

FPR Part VIII, Section D.4, and FPR Part X, Section 3.2.9, discuss smoke control and ventilation. TVA stated that plant ventilation systems at WBN are not specifically designed to exhaust smoke or corrosive gases. TVA further stated that a combination of the normal ventilation exhaust systems and portable fans are used to remove smoke from specific rooms during and after fire-fighting activities. Non-recirculating ventilation systems are provided for fire areas that may contain airborne radioactive materials. Smoke from fires that might occur in areas containing radioactive materials is monitored for radioactivity.

Based on its review of the information submitted by TVA, the NRC staff concludes that smoke control and ventilation for fire protection purposes at WBN are installed consistent with Position D.4 of Appendix A to BTP APCSB 9.5-1, and, therefore, is acceptable.

3.11   Lighting and Communications

TVA stated that fixed, self-contained lighting consisting of sealed-beam units with individual 8-hour minimum battery power supplies are provided in areas that must be manned for safe shutdown, and in access and egress routes to and from all fire areas containing equipment required for safe shutdown. TVA stated that plant walkdowns have been conducted during "blackout" conditions to assure the adequacy of the lighting. These walkdowns were used to document the adequacy of the lighting levels. Functional and visual testing of the fixed emergency lighting units is also performed to assure that the emergency lights provide their minimum 8-hour availability.

In FPR Part VII, Section 2.7, TVA requested to deviate from its emergency lighting criteria inside the reactor building, yard area, and the turbine building. Section 6.1.6 of this evaluation provides the NRC staff's evaluation of this deviation.

Based on its review of the submitted information, the NRC staff concludes, with the exception of items evaluated elsewhere in this evaluation, that the emergency lighting is consistent with the requirements in Section III.J of Appendix R to 10 CFR Part 50 and the guidelines contained in Section D.5.a of Appendix A to BTP (APCSB) 9.5-1 and therefore, is acceptable.

TVA provided several means of communication to support safe shutdown operations including (1) telephones, (2) a code, alarm, and paging system, (3) sound-powered phones, (4) cellular phones, and (5) two-way radios. The in-plant radio repeater system is the primary means of communication for performing manual shutdown actions and for fire brigade fire-fighting operations. The repeater system consists of very high frequency radio repeaters, remote control units, portable radios, and redundant antenna systems.

Operations and maintenance personnel primarily use these radios; however, the plant designates one channel of the in-plant radio system for use by the fire brigade during fires or other emergencies. Redundant fixed repeaters are widely separated so that a fire that also necessitates manual actions will not affect redundant repeaters. Some plant areas lack full radio coverage; however, coverage is available immediately outside of these rooms. Sound-powered phones are available in the ACR and local stations that are needed for alternative shutdown. Cell phones are available to supplement the communication system.

Based on its review of the submitted information, the NRC staff concludes that, with the exception of items evaluated elsewhere in this evaluation, TVA's means of communications do not take any exceptions to Positions D.5.c and D.5.d of Appendix A to BTP (APCSB) 9.5-1 and, therefore, are acceptable.

4.0 FIRE PROTECTION SYSTEMS

4.1 Water Supply and Distribution

TVA described the fire water supply system at WBN in FPR Part II, Section 12.1, "Water Supply." TVA also described the system in its response by letter dated August 5, 2011 (ADAMS Accession No. ML11224A052), to RAI FPR II-45. TVA stated that the high pressure fire protection (HPFP) water system is common to both units, and that it consists of four electric motor driven pumps and one diesel engine driven pump.

TVA stated that the electrically driven pumps are seismic Category I high-pressure vertical turbine motor-driven pumps in accordance with Section III of the American Society of Mechanical Engineers (ASME) Boiler and Pressure Vessel Code (ASME Code). Each pump is rated at 1,590 gallons per minute (gpm) at 130 pounds per square inch, gauge (psig). TVA calculated the maximum required fire water demand based on the largest automatic suppression system demand and hose streams, and it stated that each of these pumps can supply 50 percent of the required flow. The pumps are located in the seismic Category I intake pumping station (IPS) with a 3-hour-rated fire barrier that separates two fire pumps from the other two fire pumps.

TVA stated that a 100 percent capacity, UL listed, diesel fire pump is remotely located in the yard near the Unit 1 cooling tower. TVA stated that the diesel fire pump is capable of developing a flow of 2,500 gpm (100 percent capacity) at 125 psig and 3,750 gpm (150 percent capacity) at 81 psig.

TVA stated that the water supply for the electric fire pumps is taken from the Tennessee River and the diesel fire pump takes its water from the WBN Unit 1 cooling tower basin. TVA stated that the Tennessee River is essentially unlimited, and that the WBN Unit 1 cooling tower basin can provide a minimum of 2 hours supply at 150 percent of the capacity of the diesel pump.

TVA stated that the electric pumps are automatically started by activation of the fire detection systems associated with installed automatic water based suppression systems. Also, the electric pumps can be started manually from either the MCR or the appropriate 480 V shutdown board. The diesel pump automatically starts on low system pressure or can be manually started from the MCR.

TVA stated that each electric fire pump is powered from a separate 480 V shutdown board, and that in the event of loss of offsite power, each 480 V shutdown board is automatically connected to a separate emergency diesel generator. Supervised alarm circuits, indicating fire pump motor running condition and loss of line power on the line side of the switchgear, are provided in the MCR for each electric pump. The diesel fire pump also sends annunciation signals to the MCR.

TVA stated that the electric fire pumps also serve as a backup water supply to the AFW system in the event of a flood above plant grade (called "flood mode"). TVA stated that, as a result, this requires the use of pumps that meet the requirements in Section III of the ASME Code as opposed to traditional fire pump installations that are UL listed or factory-mutual-(FM)-approved pumps in accordance with NFPA 20-1993, "Standard for the Installation of Centrifugal Fire Pumps," for electric driven pumps. In FPR Part VII, Section 5.1, TVA stated the following:

(1) Pump curve verification tests have been performed to include multiple diverse points on the pump curve to replicate fire pump test requirements as opposed to the single point verification applicable to ASME Code Section III pumps;

(2) TVA performed hydraulic calculations to demonstrate that the pumps provide adequate flow and pressure to the most hydraulically remote suppression systems;

(3) The electrical circuits for pump power and control meet IEEE Class 1E standards and, even though the pumps do not start on pressure drop in the piping system, they do start on activation of the fire detection systems associated with pre-action suppression systems; and

(4) The fire pumps can only be manually stopped from the MCR or in the IPS (where the pumps are located).

Based on the above submitted information, the NRC staff concludes that, while not designed to the guidelines in NFPA 20, the electric fire pump configuration will not negatively affect the performance of the fire protection system, and meets the purpose of the guidelines of Section E.2.c of Appendix A to BTP (APCSB) 9.5-1 and, therefore, is acceptable.

TVA stated that the diesel fire pump installation and its associated controller are installed in accordance with NFPA 20-1993. Based on the information submitted by TVA that states that the diesel fire pump and associated controller are installed in accordance with NFPA 20-1993, the NRC staff concludes that the installation of the diesel fire pump is acceptable.

TVA stated that a self-cleaning strainer capable of handling 100 percent flow is provided on the discharge side of each pair of electric fire pumps. The strainers are located in the IPS and conform to the requirements in Section III of the ASME Code for seismic Category I components. For the diesel fire pump, TVA stated that mechanical screens are provided on the supply side and a strainer on the discharge.

TVA stated that the HPFP system is interconnected to the raw cooling water (RCW) system. Automatic isolation valves are provided to isolate the RCW system and selected RCW loads from the HPFP system when any fire pump is started to reduce the RCW load on the HPFP system to ensure adequate flow and pressure are available. During normal operation, HPFP system pressure is maintained by the RCW pumps.

The HPFP system mains consist of both cement lined iron yard mains and unlined steel safety headers. The steel safety headers serve as a backup water supply to the AFW system in "flood mode," as noted above. TVA stated that the details of the "flood mode" are documented in several places in the FSAR, for example, FSAR Section 2.4.14.2, "Plant Operation During Floods Above Grade." The piping inside buildings is unlined steel. The buried steel piping has an exterior coating to prevent corrosion. The electric fire pumps feed the steel headers and the diesel pump feeds the iron yard main. The two loops (iron and steel) are connected at the IPS (via normally open valve 0-FCV-26-17) and at two remote points in the auxiliary building (via normally open valves 0-FCV-26-15 and 0-FCV-26-16). TVA stated that pressure control is provided by a pressure control valve downstream of the four electric pumps.

TVA stated that sectional isolation valves are provided on the iron yard main to allow maintenance on portions of the system while the plant maintains its fire-fighting capability. In addition, TVA stated that the sectional isolation valves in the underground and building loops are locked or sealed in position and that surveillance is performed to ensure proper system alignment. The plant has not installed any sectional valves on the steel safety header. Because the two headers are redundant and because they are also connected to the iron yard main through valves in the turbine building, the plant could isolate either main and would still have two sources of fire water available.

TVA stated that all post-indicator-type valves are either sealed or locked open with a key-operated "breakaway" type lock. TVA further states that curb box valves are not locked open, but TVA considers these valves to be tamper resistant because they cannot be operated without a special "key" tool that is not generally available.

In the FPR, TVA stated that the WBN fire water supply system as being able to provide the designed fire-fighting capacity either with one electric pump and the diesel pump unavailable or with the hydraulically least demanding portion of any loop main out of service. TVA further stated that the design flow demand consists of design flow to the largest sprinkler or water spray system plus design flow to non-isolated RCW loads and 500 gpm for hose streams.

TVA stated that suppression systems and hose station standpipe systems are separately connected to the yard main or to headers within buildings and are fed from each end of the building, so that a single failure cannot impair both suppression systems and hose stations at the same time.

As result of the concern with microbiologically induced corrosion (MIC) and other corrosion issues, TVA has instituted a permanent monitoring program for assuring the performance of the standpipe and suppression systems. TVA stated that this testing is performed at the hydraulically most remote hose stations every 3 years. TVA uses the calculated design basis pressure and flow requirements for these hose stations as the basis to monitor system performance.

TVA's design calculation reduces the actual pipe inside diameter by 0.8 inches and uses a Hazen-Williams C factor of 55 for the sections of piping that are normally wetted. TVA stated that the purpose of these piping restrictions and the C factor of 55 are to account for the 40-year service life of the pipe. The data collected from these tests will be compared to the calculated values and trended to detect system failure.

TVA stated that all raw water systems, including HPFP, are chemically treated in a manner that is consistent with nuclear industry practice. TVA stated that this treatment includes oxidizing biocide, non-oxidizing biocide, phosphate, and zinc. TVA further stated that the oxidizing and non-oxidizing biocides are used to control Asiatic clams, zebra mussels, slime, and MIC; the phosphate is used to sequester iron from existing corrosion products; and the zinc acts as a mild corrosion inhibitor for the carbon steel surfaces. As described in TVA's letter dated August 5, 2011 (ADAMS Accession No. ML11224A052), in response to RAI FPR VII-2.1, the non-oxidizing biocide treatments are coordinated with periodic system flushes in order to distribute the biocide to normally stagnant portions of the system.

TVA stated that two programs have been implemented to combat pipe corrosion. First, TVA implemented the Corrosion Control Program, which primarily monitors pipe wall thickness using ultrasonic techniques, replacing lengths of pipe when minimum wall thickness cannot be maintained. Additionally, TVA stated that a WBN Buried Piping Plan has been established in support of NEI 09-14, "Guideline for the Management of Underground Piping and Tank Integrity." TVA described this program as providing for the risk ranking of buried piping relative to installed conditions (e.g., design and construction practices, as well as soil characteristics) and consequences of a failure of the piping. TVA stated that these programs are intended to provide assurance in the integrity of the HPFP system boundaries.

In addition, TVA performed a code compliance review against NFPA 24-1973, "Outside Protection," as documented in FPR Part X. No substantial exceptions were identified.

Based on its review of the submitted information, the NRC staff concludes that, with the exception of the system design demand, the fire water supply system conforms to the guidelines of Sections E.2 and E.3.a of Appendix A to BTP (APCSB) 9.5-1 and, therefore, is acceptable. Regarding the system design demand, the NRC staff concludes that it conforms to the NRC's current guidance found in Position 3.2.1 of RG 1.189, Revision 2, and, therefore, is acceptable.

4.2   Active Fire Control and Suppression Features

4.2.1   Automatic Fire Suppression Systems

4.2.1.1   Sprinklers and Fixed Spray Systems with Closed Heads

In FPR Part III, Section 10.3.1, TVA stated that all AVs that contain redundant safe shutdown equipment are protected to ensure that the plant maintains its safe shutdown capability. In most cases, the means of protection is consistent with Section III.G.2 of Appendix R to 10 CFR Part 50. For instance, in areas in which cables of redundant safe shutdown equipment are located in an AV and could be damaged by fire, the plant ensures the function by installing a 1-hour ERFBS on one train with automatic fire detection and suppression or by installing a 3-hour ERFBS in areas that do not provide fire detection and suppression. TVA also stated that, if separation between rooms in the same fire area is less than 3 hours rated, the plant either provides automatic detection and suppression systems or identifies and justifies deviations.

Where deviations from Section III.G.2 requirements exist, with respect to coverage of suppression and detection systems, TVA performed evaluations to demonstrate that an adequate level of protection is provided. Section 6.1.7 of this evaluation provides the NRC staff evaluation of these deviations.

In FPR Part VII, Section 2.4, TVA described a methodology used to resolve situations where redundant trains are separated by more than 20 feet, but without 20 feet free of intervening combustibles. Section 6.1.4 of this evaluation provides the NRC staff evaluation of this configuration.

Where provided, TVA stated that sprinkler systems and fixed water spray systems are designed in accordance with the applicable requirements in NFPA 13-1975, "Standard for Installation of Sprinkler Systems," and NFPA 15-1973, "Standard for Water Spray Fixed Systems."

In addition, TVA performed a code compliance review and identified several areas in which the sprinkler and fixed spray systems differed from the code. The important exceptions to the NFPA 13-1975 code identified were as follows:

- Fire department pumper connections for the sprinkler system are only provided to buildings with one connection to the underground fire main. (NFPA 13, Section 2-7). The NRC staff concludes that this arrangement meets the intent of the provision.

- Strainers are provided in the supply to each pre-action sprinkler system in lieu of following flushing requirements. (NFPA 13, Section 3-37.3.) The NRC staff concludes that this arrangement meets the intent of the provision.

- Sprinklers are not provided below the open grating above the high-pressure fire pump flow control valve on elevation 692 feet in the Unit 1 penetration room (Room 692.0-A7). TVA stated that the combustible loading in this fire area is insignificant. This grating is approximately 5 feet wide by 15 feet long and is 15 feet above the room floor. Two sprinklers are installed approximately 3 feet above the grating. Plant procedures prohibit the storage of material on these grated walkways, so the gratings would be free of foreign obstructions. Due to the size of the grating (4 in. by 1 in.), flow from the sprinklers is not expected to be restricted by the grating. (NFPA 13, Section 4-4.11.) The NRC staff concludes the current sprinkler configuration in the Unit 1 penetration room is acceptable.

The NRC staff reviewed the other code exceptions from NFPA 13 that TVA proposed in FPR Section X, and determined that the exceptions will not affect the performance of the systems and, therefore, are acceptable.

With respect to NFPA 15, TVA did not take any exceptions to the code for the water spray systems protecting outdoor transformers, the hydrogen trailer, turbine hydrogen seal oil unit, or the turbine lube oil reservoir. TVA used the guidance of NFPA 13 to design the directional fusible nozzle water spray systems used to protect certain charcoal filters and the RCPs.

TVA stated that automatic pre-action sprinklers are provided in areas in which it is important to prevent accidental discharge of water. Operation of the pre-action sprinkler system is initiated by a signal from the fire detection system in the area. Actuation can also be initiated manually by mechanical operation at the deluge valve. In addition, selected pre-action systems at WBN have manual actuation stations placed at strategic locations remote from the valve. These systems are provided with air supervision if the piping downstream of the system control valve supplies more than 20 sprinkler heads.

TVA stated that, where manually activated suppression systems are installed, the piping network isolation valve is maintained in the closed position. Personnel are alerted to a problem in these areas by the fire detection system and, after confirming there is a fire, personnel open the appropriate isolation valve to allow water into the system. Water is then applied to the fire when the heat from the fire melts the fusible element in the sprinkler head. Water flow is subsequently stopped by manually closing the associated isolation valve.

In FPR, Part VIII, TVA stated that drainage is provided to remove the expected fire protection water flows or control the accumulation of water such that the water will not cause unacceptable damage to equipment in the area. TVA further stated that additional drainage can be achieved by diverting water into adjacent rooms. Finally, TVA stated that water draining from areas which may contain radioactivity is sampled and analyzed before being discharged into the environment.

TVA stated that standpipes, hose stations, and portable fire extinguishers are provided throughout the control building, but that fixed fire suppression systems are not provided for all rooms. TVA justified the lack of fixed automatic suppression capability by stating that the control building is a single fire area with fire detection provided throughout the control building except in certain areas, and that there are no alternative shutdown cables or equipment located in the control building, thereby satisfying the design intent of maintaining safe shutdown capability for a postulated fire event by providing an alternate design concept. Based on TVA's justification, the NRC staff concludes TVA's approach to be acceptable. See Section 3.3 of this evaluation for a discussion of alternate shutdown, and Section 6.1.3 for a detailed discussion of the lack of area-wide automatic suppression in the control building.

In all cases, TVA stated that an adequate level of protection is provided via a combination of limited combustible materials, administrative controls, fire rated barriers, spatial separation, and active fire protection systems. Where exceptions or deviations from NRC staff guidance, rules, or design standards exist, TVA stated that they have performed evaluations to ensure that an adequate level of protection is provided. The NRC staff reviewed TVAs approach to the use of sprinkler and water spray fire suppression systems, and concludes that TVA's design criteria and bases are consistent with Positions E.2 and E.3.c of Appendix A to BTP (APCSB) 9.5-1 and the defense-in-depth concept described in Appendix R to 10 CFR Part 50, and therefore, are acceptable.

4.2.1.2 Gas Suppression Systems

TVA stated in FPR Part II that automatic total flooding CO_2 suppression systems are provided for the auxiliary instrument rooms and computer room in the control building; and in the lube oil storage room, diesel engine rooms (4), fuel oil transfer room, and 480 V board rooms (4), located in the diesel generator building.

TVA stated that the CO_2 systems are designed and installed in accordance with the NFPA 12-1973, "Carbon Dioxide Extinguishment Systems," the code of record for these systems. Further, TVA stated in its letter dated March 16, 2011 (ADAMS Accession No. ML13060A403), in response to NRC question RAI FPR II-6, that the systems installed in the computer room, diesel generator electrical board rooms, lube oil storage room, and fuel oil transfer room are installed for property protection purposes only, and do not have soak time requirements. In addition, TVA stated that the systems are appropriate for the anticipated hazards and that they performed system dump tests to ensure agent concentration, agent reserve, and operability of the distribution system.

TVA stated that a signal from either the fire detection system or a push button station activates the area alarms, CO_2 discharge timer, which actuates the master control valve, and the area selector valve permitting the CO_2 to be discharged into the selected area. In addition, the system can be manually operated via the electro-manual pilot valve for each hazard protected on the loss of power to the system. In designing these systems, TVA has considered personnel safety by providing the pre-discharge alarm to notify anyone in the area that CO_2 is going to be discharged, and by adding an odorant to the CO_2 to warn personnel that the system has been discharged.

In addition, TVA stated that the actuation of these systems causes selected fire dampers and doors to the protected area to close and the HVAC fans to the area to shut down, ensuring that the minimum concentration of CO_2 is maintained and preventing fire spread from the area of fire origin. TVA also stated that it has performed full discharge tests for representative rooms in conjunction with door fan pressurization tests to validate CO_2 concentration and soak times.

The CO_2 storage tank for supplying CO_2 to systems that protect the diesel generator building is located in the diesel generator building. The diesel generators are protected from the effects of a postulated failure of this tank by an 18-inch thick reinforced concrete wall. The vent path for the tank room for the storage tank compartment is through a set of double doors that lead into a stairwell then, if needed, through another set of double doors which open to the atmosphere from the stairwell.

The CO_2 for the balance of the plant is supplied from a storage tank in an underground vault in the yard. TVA stated that the system is designed such that failure of the system cannot pose a threat to any safety-related areas or structures.

The NRC staff has reviewed TVA's approach to the use of automatic CO_2 fire suppression systems and concludes that TVA's design criteria and bases are consistent with Positions D.4.i and E.5 of Appendix A to BTP (APCSB) 9.5-1 and, therefore, are acceptable.

4.2.2   Manual Suppression Capability

4.2.2.1   Hose Stations

In FPR, Part II, TVA stated that hose stations for manual fire-fighting are located throughout the plant to ensure that an effective hose stream can be directed to any safety-related area in the plant. TVA further stated that the system is designed according to the requirements of NFPA 14-1974, "Standpipe and Hose System for Sizing, Spacing, and Pipe Support Requirements," except for those hose stations in certain areas of the plant in which TVA requested a deviation to exceed the 100-foot hose spacing limitation. These deviations are discussed in Section 6.2.4 of this evaluation.

In addition, TVA performed a code compliance review and identified several areas in which the manual fire-fighting hose stations and standpipe system differed from the code. TVA also performed evaluations to justify these exceptions. The significant NFPA 14 code exceptions identified and associated justifications are:

- The standpipes located on elevations 676.0 feet, 692.0 feet, 713.0 feet, 729.0 feet, 757.0 feet, 772.0 feet, and 782.0 feet of the auxiliary building are supplied by a 3-inch pipe rather than the 4-inch pipe, and elevation 755.0 feet of the control building has 2 ½-inch supply piping. TVA stated that it verified by hydraulic calculation that these pipe sizes were adequate. (NFPA 14, Section 212.)

- Two standpipes (0-26-677 and -690) are not provided with header isolation valves. TVA stated that these systems can be isolated and that this would not preclude the ability to provide hose stream coverage in the locations normally served by these standpipes. (NFPA 14, Sections 413 and 622.)

- Pressure reducing devices are not installed at the hose stations. TVA justified this by stating that the hose stations are for fire brigade use, and the fire brigade personnel are trained in the use of high pressure fire hoses. TVA further stated that the hoses and related fittings are maintained to accommodate the expected system pressures. (NFPA 14, Section 442.)

- High pressure valves, pipes, and fittings are not used, even though system spikes of up to 190 psi occur due to pump start surges. TVA stated that the piping and fittings can withstand the working pressures of the system and that the system is in accordance with American National Standards Institute B31.1, "Code for Pressure Piping," system requirements. (NFPA 14, Sections 625, 631, and 641.)

- Water flow alarms are not provided on all standpipes. TVA stated that the hose stations are provided for fire brigade use. Other site personnel are trained to report fires before using fire-fighting equipment (if they have been trained in its use). Therefore, TVA concluded that sufficient notification of standpipe use will be provided to the MCR without water flow alarms. (NFPA 14, Section 67.)

The NRC staff reviewed TVA's proposed exceptions from NFPA 14 and determined that they will not affect the performance of the hose stations and the standpipes. Therefore, the exceptions are acceptable.

In FPR, Part VIII, TVA stated that drainage is provided to remove the expected fire protection water flows or control the accumulation of water such that the water will not cause unacceptable damage to equipment in the area. TVA further stated that additional drainage can be achieved by diverting water into adjacent rooms. Finally, TVA stated that water draining from areas that may contain radioactivity is sampled and analyzed before being discharged into the environment.

TVA stated in the FPR that hose station nozzles appropriate for the expected hazards (e.g., electrically safe) are provided for each hose station. In addition, TVA stated, in FPR Part VIII, and in its letter dated August 5, 2011 (ADAMS Accession No. ML11227A257), in response to RAI FPR II-41.1 and RAI FPR VII-17.1, that provisions has been made to supply water at sufficient pressure and capacity to the standpipes, hose stations, and hose connections for manual fire-fighting in areas required for safe plant shutdown in the event of a safe-shutdown earthquake.

Based its review of the information submitted by TVA, the NRC staff concludes, with the exception of items evaluated elsewhere in this evaluation, that the standpipe system and hose stations do not take any exceptions to Positions E.3.d and E.3.e of Appendix A to BTP (APCSB) 9.5-1 and, therefore, are acceptable.

4.2.2.2   Fire Extinguishers

TVA stated that portable fire extinguishers of a size and type compatible with specific hazards are strategically located throughout the plant for use by trained personnel. TVA also stated that fire brigade members and fire watch personnel have been trained on the location of extinguishers for firefighting operations through the extinguishers inspection program. In addition, TVA stated that fire extinguishers are inspected on a quarterly basis.

TVA's proposed application and strategic distribution of portable fire extinguishers throughout the plant is consistent with the guidance contained in Position E.6 of Appendix A to BTP (APCSB) 9.5-1, and provides reasonable assurance that fire extinguishers will be readily available and quickly accessed in the event of a fire emergency. Therefore, the NRC staff concludes that TVA's proposed application and strategic distribution of portable fire extinguishers is acceptable.

4.3   Fire Detection Capability

In FPR Part III, Section 10.3.1, TVA stated that all analysis volumes containing redundant safe shutdown equipment are protected to ensure safe shutdown capability is maintained. In most cases, the means of protection is consistent with 10 CFR Part 50, Appendix R, Section III.G.2. For instance, where cables of redundant safe shutdown equipment is located in an analysis volume and could be damaged by fire, the function is ensured either by the installation of 1-hour ERFBS on one train with automatic fire detection and suppression, or a 3-hour ERFBS where fire detection and suppression are not provided. TVA also stated that if separation between rooms in the same fire area is less than 3-hour rated, automatic detection and suppression systems are provided or deviations are identified and justified.

Where deviations from Section III.G.2 requirements exist, with respect to coverage of suppression and detection systems, TVA performed evaluations to demonstrate that an adequate level of protection is provided. Section 6.1.7 of this evaluation provides the NRC staff evaluation of these deviations.

In FPR Part VII, Section 2.4, "Intervening Combustibles," TVA described a methodology used to resolve situations where redundant trains are separated by more than 20 feet, but without 20 feet free of intervening combustibles. Section 6.1.4 of this evaluation provides the NRC staff evaluation of this configuration.

As described by TVA in FPR Part II, the fire detection system consists of initiating devices, local control panels, a remote transmitter-receiver providing a remote multiples function, computerized multiplex central control equipment, and a power supply. A central processor unit (CPU) of the computerized multiplex central control equipment communicates with the local control panels via the remote transmitter/receiver units over looped circuits. TVA stated that where detection is provided for the protection of safety-related or FSSD equipment, Class A, four-wire, supervised circuits link the fire detectors to the local control panels and annunciate status change to a constantly attended location. In addition, a second CPU is provided in a constantly attended location as an alternate for the primary processor.

TVA stated that the fire detection system uses photoelectric, ionization, and thermal detectors. The fire detection system also monitors duct detectors and devices for monitoring fire suppression system piping integrity, water or CO_2 flow, and diesel fire pump status. The fire detection system gives an audible and visual alarm, and also annunciates in the control room.

TVA stated that, where detection systems are provided, the detection systems are designed in accordance with the applicable requirements of the NFPA 72D-1975, "Installation, Maintenance and Use of Proprietary Signaling Systems," and NFPA 72E-1974, "Automatic Fire Detectors." In addition, TVA performed a code compliance review and identified several areas in which the systems differed from the code. The significant NFPA 72D and NFPA 72E code exceptions identified were as follows:

(a) The operation and supervision of fire alarms is not the primary function of the system operators (i.e., the control room operators). The operators are responsible for all control room alarms and controlling the plant. (NFPA 72D, Section 1223.) This is acceptable to the NRC staff, because, consistent with the role and training of the operators, a fire alarm actuation is an event that will be responded to, and will not be ignored.

(b) The fire alarm console in the MCR is an UL-listed device; however, TVA modified this console by adding non-UL-listed panels known as A-B switchover panels, which allow a quick changeover to the installed spare control system. TVA stated that this option is not commercially available and does not degrade the system. The two alerting tone volume control devices have been adjusted to meet the requirements of the human factors analysis for the MCR. (NFPA 72D, Sections 1213 and 2022.)

The NRC staff concludes that this modification is acceptable because it does not diminish the ability of the system to perform its function.

(c) Actions upon receipt of a fire alarm signal; the fire brigade is not immediately notified. TVA stated that, upon receipt of an alarm from a detection system, an individual (auxiliary or fire operator) is dispatched to the area to determine the cause of the alarm. If a fire exists, the individual notifies the MCR, and control room operators notify the plant fire brigade. If both detection zones alarm of a cross-zoned detection system, the fire brigade is notified immediately. (NFPA 72D, Section 1251.) The NRC staff concludes that this arrangement is acceptable because it allows false alarms to be addressed while maintaining rapid response by the site fire brigade to actual fires.

(d) The fire alarm system uses the emergency diesel generators as the automatic secondary power supply. The uninterruptible power supply backup and batteries inside the fire alarm console supply selected devices within the console. (NFPA 72D, Sections 2223, 2224, and 2231.) The NRC staff concludes that this arrangement is acceptable because it provides a reliable source of backup electrical power.

(e) Signal attachments and circuits (pressure switches) can be removed or tampered with and not cause an alarm. The site personnel access control system and the work control system provide assurances that work on such devices is properly controlled and documented. (NFPA 72D, Section 3423.) The NRC staff concludes that this is acceptable because these devices are not installed in areas accessible to the general public, where tampering is a concern.

(f) Sprinkler system control valves are not electrically supervised; instead, the valves are locked open or sealed open and periodically inspected instead. TVA stated that administrative controls, including second party verification of position and strict site-access and work controls, will ensure that valves are in the correct position. (NFPA 72D, Section 3442.) The NRC staff concludes that this is acceptable because it provides assurance that the valves will be in the correct positions when needed.

(g) Both visual and recorded displays meet the code, but records are not preserved for later inspection. Plant procedures have reporting requirements for conditions adverse to quality. These procedures require an adverse condition report to be completed before the end of the shift on which the problem was identified. Documentation from the fire alarm printout would be available to support the adverse condition report. (NFPA 72D, Section 4111.) The NRC staff concludes that this arrangement is acceptable because it will support the reconstruction of the sequence of events.

(h) The transmission of an alarm signal to the fire alarm console from a wire-to-wire short circuit cannot be recorded. TVA stated that a wire-to-wire short will generate a trouble signal which requires corrective action and associated compensatory measures as laid out in FPR Part II, Section 14. (NFPA 72D, Sections 4112 and 4311.) The NRC staff concludes that this is acceptable because this situation initiates corrective actions and compensatory measures, which include roving or continuous fire watches.

(i) Fire detection has not been provided in the diesel generator building stairway D1, bathroom, and CO_2 storage room on elevation 742 feet, and the corridor and radiation shelter room on elevation 760 feet, because a fire in these rooms would not impact the plant's ability to achieve and maintain safe shutdown. In addition, no detection capability is installed in specific auxiliary building pump room entrance labyrinths, the airlocks, and the auxiliary building elevator shaft and associated auxiliary elevator equipment. TVA stated that a fire in these areas would not have an impact on the safe shutdown capability of the plant. (NFPA 72E, Section 2-6.5.) The NRC staff concludes that this is acceptable because of the nature of the spaces (e.g., the lack of combustibles, lack of impact on safe shutdown capability.)

(j) Smoke detectors in the high ceiling areas of the plant are not installed alternately on two levels. TVA has addressed the issue of high ceilings by reducing the spacing of the detectors at the ceiling level. This reduced spacing is used on auxiliary building elevations 692 feet, 713 feet, 737 feet, 757 feet, and the waste packaging room. (NFPA 72E, Section 4-4.5.2.) The NRC staff concludes that this is acceptable because stratification is not a concern due to the ventilation system.

(k) TVA uses duct detectors in lieu of area detectors in the reactor building upper and lower compartment coolers to provide protection specifically for the coolers. TVA stated that the regulatory requirements for detectors are met for the remainder of the reactor building. (NFPA-72E, Section 8-1.1.2.) The NRC staff concludes that this is acceptable because these detectors are installed to protect these specific pieces of equipment (e.g., the compartment coolers) and not the general area.

(l) Duct detectors are not provided per NFPA 90A requirements, which require that activation of a detector automatically stops the ventilation system. Instead, fans serving the area of the plant containing the fire are shut down manually to ensure that air flow will not prevent fire dampers from closing. (NFPA-72E, Section 8-1.2.1.) The NRC staff concludes that this is acceptable because it accomplishes the goal of the provision. Additionally, the HVAC system has been designed as described in WBN Final Safety Analysis Report (FSAR) chapters 3, 6, and 9, and approved by the NRC.

The NRC staff has reviewed TVA's proposed exceptions from NFPA 72D and NFPA 72E, and has determined that they will not affect the performance of the affected systems or the ability of the plant to achieve and maintain safe shutdown. Therefore, the exceptions are acceptable. The NRC staff concludes that TVA's design criteria and bases for the installed systems are consistent with Position E.1 of Appendix A to BTP (APCSB) 9.5-1 and, therefore, are acceptable.

In all cases, TVA stated that an adequate level of protection is provided via a combination of limited combustible materials, administrative controls, fire rated barriers, spatial separation, and active fire protection systems. Where exceptions or deviations from NRC staff guidance, rules, or design standards exist, TVA stated that they have performed evaluations to ensure that an adequate level of protection is provided. The NRC staff reviewed TVA's approach to the use of fire detection systems and concludes that, with the exception of items evaluated elsewhere in this evaluation, TVA's design criteria and bases are consistent with Position E.1 of Appendix A to BTP (APCSB) 9.5-1 and the defense-in-depth concept described in 10 CFR Part 50, Appendix R, and, therefore, are acceptable.

5.0 FIRE PROTECTION FOR SPECIFIC PLANT AREAS AND HAZARDS

5.1 Containment

Appendix A to BTP (APCSB) 9.5-1, includes guidance for fire protection in containment. In its letter dated May 26, 1995 (ADAMS Accession No. ML073230888), TVA stated that a major fire hazard within containment is the lube oil in the RCPs. If oil leaks from the RCPs, an oil collection system is available to collect the oil for each RCP as described below. This system on each RCP is designed to collect oil from all potential leakage locations, including the RCP oil lift pump, system piping, overflow lines, the lube oil cooler, oil fill and drain lines, flanged connections on the oil lines, and the lube oil reservoirs.

The RCPs, lubricating oil systems, oil spray shields, oil collection basins, drain piping, and containment sump are designed to seismic Category I requirements so that they will not fail during a safe shutdown earthquake.

Each of the four RCPs is protected by an automatic fire suppression and detection system. A heat collection hood is installed directly above the RCP motors. In the event of an RCP motor fire, the heat collection hood acts as a ceiling, that forces the heat to stall around the detectors and the suppression nozzles, thus reducing the response time of these fire protection devices.

Section 6.1.8 of this evaluation provides the NRC staff's evaluation of the RCP oil collection system configuration and associated fire protection features.

TVA stated that areas of divisional interaction within the annulus areas are protected by automatic fixed water-spray systems and ionization smoke detectors. Additionally, fixed water-spray systems are provided for the charcoal and HEPA filters in the lower containment air-cleanup units. Thermal detectors are provided for the charcoal filters and HEPA filters. Ionization duct detectors are provided for each lower containment cooling unit and each upper compartment cooling unit. In addition, ionization smoke detectors are provided for the exhaust ducts serving the containment purge and air exhaust systems and the emergency gas treatment system.

TVA stated that a standpipe and hose system is provided in each containment to complement the installed automatic suppression systems. The standpipe systems are normally dry and admit water when a remote control device installed at each hose station is manually operated.

TVA stated that RESs are relied on to separate of cables and associated non-safety circuits of redundant trains. TVA evaluated the combustibility of the RES material in FPR Part VII, Section 2.2, "Non-Combustible Radiant Energy Shields." Section 6.1.2 of this evaluation provides the NRC staff's evaluation of this configuration.

TVA stated that the RCP oil collection system meets the requirements in Section III.O, "Oil collection system for reactor coolant pump," of Appendix R to 10 CFR Part 50 with the exception of a deviation to allow for minor amounts of oil that become entrained in the ventilation air to escape the oil collection system. TVA evaluated the RCP oil collection system in FRP Part VII, Section 2.8, "Reactor Coolant Pump Oil Collection System." See Section 6.1.8 of this evaluation for a detailed evaluation of the deviation.

Based on its review of the information provided by TVA, with the exception of items evaluated elsewhere in this evaluation, the NRC staff concludes that the fire protection features for containment conform to the guidance in Position F.1 of Appendix A, to BTP (APCSB) 9.5-1, and, therefore, are acceptable.

5.2   Control Room Complex

5.2.1   Control Room

Appendix A to BTP (APCSB) 9.5-1 includes guidance for fire protection in the MCR. The MCR is common to both units and contains circuits for safe shutdown for fires outside of the control building. TVA designated the control building, which contains the MCR, an alternative shutdown area. As a result, independent alternative shutdown capability has been provided for this area. Discussion of alternative shutdown is located in FPR Part IV and Section 3.3 of this evaluation. The entire control building is considered a single fire area and is separated from other fire areas (e.g., the auxiliary building, turbine building) by 3-hour fire barriers, as documented in FPR Part VI.

In FPR Part VII, Section 2.6.4, TVA evaluated the affect of nonrated metal hatch covers between the mechanical equipment rooms and the turbine building. Section 6.2.7.4 of this evaluation provides the NRC staff's evaluation of this deviation.

FPR Part VIII summarizes the fire barriers that separate the MCR from the balance of the control building. The MCR is separated from adjacent rooms on the same elevation in the control building by 1-hour rated fire barriers. Doors between the control room and the turbine building and the control room and auxiliary building are 3-hour fire-rated doors. The MCR and the cable spreading room are not separated by a rated fire barrier.

FPR Part VIII describes the use of cables in the MCR. TVA stated that (1) wiring for lighting terminates in the lighting fixtures, (2) instrumentation and control wiring enters through the bottom of cabinets and runs only inside the panels or control boards in which the wires are terminated, and (3) cable are not routed through the control room from one area to another area.

In FPR Part VIII, TVA described manual fire-fighting operations. TVA stated that fire extinguishers are provided in the MCR. Standpipe and hose stations are located in the stairwells at each end of the MCR. TVA also stated that the hose stations have electrically qualified nozzles in alignment with the expected hazards.

TVA stated that ionization smoke detectors are provided in selected cabinets, and additional ionization detectors are installed in the MCR ventilation system. TVA further stated that fire alarms in other parts of the plant, as well as the MCR, alarm and annunciate in a constantly attended location in the MCR.

FPR Part VIII also summarizes smoke control features for the MCR. The MCR ventilation air intakes are provided with remotely controlled dampers to prevent smoke from entering the control room. Manual venting of the control room can be achieved by using portable smoke ejectors available onsite and by opening the doors of the MCR. TVA also stated that breathing apparatuses are available for the control room NRC staff.

TVA evaluated the impact of not providing an automatic suppression system (as required for alternative shutdown locations) in the MCR and corridor in FPR Part VII, Section 2.3. Section 6.1.3 of this evaluation provides the NRC staff's evaluation of this deviation.

In FPR Part VII, Section 4.1, TVA evaluated MCR Doors C49 and C50 for altering the doors by adding signs and security hardware or by repairing onsite damage. Section 6.2.2 of this evaluation provides the NRC staff's evaluation of this configuration.

Based its review of the information submitted by TVA, the NRC staff concludes that, with the exception of items evaluated elsewhere in this evaluation, an equivalent level of safety to the separation requirements in Position F.2 of Appendix A to BTP (APCSB) 9.5-1 has been achieved by TVA because of (1) the installed detection and suppression in the cable spreading room, (2) the low combustible loading and installed automatic suppression and detection in adjacent non-MCR control building areas, (3) the provision for alternative shutdown for control building fires through use of the independent ACR complex, and (4) the FSSD evaluation that demonstrates the use of the ACR to achieve post-FSSD, and therefore, is acceptable.

5.2.2   Auxiliary Control Room

TVA designated the control building as an alternative shutdown area. FSSD activities take place outside of the control building for large or damaging fires in the control building. The ACR at WBN provides independent alternative shutdown capability for control building fires. Discussion of alternative shutdown is located in FPR Part IV and Section 3.3 of this evaluation.

TVA stated that the ACR is independent from the control building, which includes the cable spreading room, MCR, and auxiliary instrument room. The ACR is located in the auxiliary building, and is divided into five independent, dedicated rooms. Each room is separated from the others and from the rest of the auxiliary building by at least 2-hour rated fire barriers and from the control building by 3-hour rated fire barriers. The five independent rooms consist of a Train A and a Train B transfer switch room for each unit and a common ACR containing multiple instrumentation and control panels for both units. Ionization smoke detectors and pre-action sprinkler system are provided in each of the five rooms. Standpipe and hose stations are provided for manual fire-fighting activities in the ACR complex from adjacent Rooms 757.0-A2 and -A24.

In FPR Part IV, TVA described the ACR as designed to control the FSSD activities after control has been established at the ACR following MCR abandonment. Systems requiring operator manipulations have the controls located in the ACR along with their associated transfer switches located in the adjacent transfer switch rooms. TVA stated that operators are periodically trained in shutdown procedures from the ACR. TVA further stated that the instruments and controls located in the ACR are separated from, or can be electrically isolated from, the corresponding instrumentation and controls located in the control building.

In FPR Part VII, Section 2.4, TVA evaluated the affect for intervening combustibles, such as insulation on cables in trays and Thermo-Lag® in the ACR. Section 6.1.4 of this evaluation provides the NRC staff's evaluation of this deviation.

Based on its review of the information submitted by TVA, the NRC staff concludes that, with the exception of items evaluated elsewhere in this evaluation, the installed fire protection features are consistent with the NRC's current guidance in Position 6.1.6 "Alternative and Dedicated Shutdown Panels," in Revision 2 to RG 1.189, and therefore, are acceptable.

5.3   Cable Spreading Room

The cable spreading room is common to both units and contains circuits for redundant safe shutdown features. TVA designated the control building, which contains the cable spreading room, an alternative shutdown area. As a result, independent alternative shutdown capability has been provided for this area. Discussion of alternative shutdown is located in Part IV of the FPR and Section 3.3 of this evaluation.

TVA stated that the cable spreading room is separated from the adjacent buildings by 3-hour rated barriers. TVA also stated that fire brigade access to the cable spreading room is provided by doors from the turbine building and from enclosed stairways within the control building. TVA stated that portable extinguishers which are located inside and immediately outside the cable spreading room are available. Additionally, standpipe and hose stations are provided from the two stairwells and from the turbine building.

In the FPR Part VIII, TVA summarized the fire protection features for the cable spreading room and stated that these features provide full coverage detection and automatic suppression. The automatic pre-action sprinkler system has a ceiling layer and an intermediate layer of sprinklers under the grating and staggered between the upper level heads. TVA further stated that the installed cables are designed to allow wetting without faulting.

Based on its review of the information submitted by TVA, the NRC staff concludes that the fire protection features for the cable spreading room do not take any exceptions to Position F.3 of Appendix A to BTP (APCSB) 9.5-1 and, therefore, are acceptable.

5.4   Switchgear Rooms

TVA stated that the Trains A and B 6.9 kV and 480 V switchgear rooms are located within the auxiliary building, but separated from each other and from other rooms within the auxiliary building by 2-hour fire rated barriers and from the control building by 3-hour fire rated barriers. Each room is provided with a full area coverage automatic pre-action sprinkler system that is actuated by a cross-zoned area-wide ionization smoke detection system. Water spray shields have been installed to protect safety related electrical equipment against the effects of inadvertent or advertent actuation of the automatic suppression system. Additionally, standpipe and hose stations are provided in each of the switchgear rooms.

Based on its review of the information submitted by TVA, the NRC staff concludes that the fire protection features for the essential switchgear rooms provide an equivalent level of fire safety to Position F.5 of Appendix A to BTP (APCSB) 9.5-1 and, therefore, are acceptable.

5.5   Battery Rooms

TVA stated in FPR Part VIII, that the required Vital Battery Rooms I through IV are separated from all other plant areas by 3-hour fire rated barriers. The Fifth Vital Battery Room is a spare that can be used for any of the other four vital batteries. TVA further stated that the Fifth Vital Battery Room is separated from other plant areas by 2-hour fire rated barriers that exceed the hazards to which they could be exposed.

TVA also stated that ceiling vents are provided for each battery room with a direct exhaust to outside the building to maintain the concentration of hydrogen below 2 percent by volume within the battery rooms. Additional details of these exhaust systems are available in WBN Unit 2 FSAR Section 9.4.3.2.5, "Auxiliary Board Rooms Air-Conditioning Systems."

TVA provided a summary of the fire protection features for the battery rooms in FPR Part VIII. TVA stated that full coverage automatic smoke detection and manually actuated sprinkler system are provided for Vital Battery Rooms I to IV. Smoke detection and an automatic pre-action sprinkler system are provided for the Fifth Vital Battery Room. With regard to manual fire-fighting, TVA stated that hose stations and portable fire extinguishers are available for fire brigade use.

Based on its review of the information submitted by TVA, the NRC staff concludes that the fire protection features for the battery rooms do not take exceptions to Position F.7 of Appendix A to BTP (APCSB) 9.5-1 and, therefore, are acceptable.

5.6   Turbine Lubrication and Control Oil Storage and Use Areas

TVA stated in FPR Part VI that a fire in the turbine building would not impact equipment required to achieve safe shutdown, and that Train A and B systems and components would be utilized without mitigating actions. TVA further stated that cable tray penetrations through the 3-hour fire rated fire barrier separating the turbine building from the control building are sealed with 3-hour fire-rated penetration seals and are provided with automatic water curtain protection on the turbine building side. TVA stated in FPR Part VIII that turbine building oil hazards are protected by fixed water spray systems. Additionally, standpipe and hose stations are provided on each elevation of the turbine building.

Based on its review of the information submitted by TVA, the NRC staff concludes that the fire protection features for the turbine building provide an equivalent level of safety as the guidelines in Position F.8 of Appendix A to BTP (APCSB) 9.5-1 and, therefore, are acceptable.

5.7   Diesel Generator Areas

In the FPR Part VIII, TVA stated that the diesel generator building is remotely located and is not adjacent to any other safety related building or structure, and that each diesel generator and its associated equipment are separated from each other by 3-hour fire barriers.

TVA described the automatic fire suppression systems installed in these areas as follows. Each diesel generator area is provided with full coverage detection that alarms and annunciates in the control room and alarms locally. Automatic, total flooding CO_2 suppression systems protect each diesel generator, the associated day tanks, and the electrical board room. TVA also stated that the diesel generator building pipe gallery and corridor are protected by a pre-action sprinkler system. For manual suppression, TVA stated that standpipes and hose stations are available on both elevations of the diesel generator building, with back-up from hydrants in the yard.

TVA stated that the two 550-gallon day tanks are located in the same room as the associated tandem diesel generator.

Based on its review of the information submitted by TVA, the NRC staff concludes that the fire protection features for the diesel generator areas do not take any exceptions to Position F.9 of Appendix A to BTP (APCSB) 9.5-1 and, therefore, are acceptable.

5.8   Diesel Generator Fuel Oil Storage Areas

In FPR Part VIII, TVA stated that the above ground diesel fuel oil storage tanks are located more than 50 feet from any safety related building or structure, and that they are located within a diked area sized to contain leaks or spills of fuel oil.

TVA further stated that the 7-day fuel oil storage tanks for each diesel generator are buried under the floor of the diesel generator building. The only portions of the tanks that are not buried are the manway access openings to each tank within the diesel rooms and in the common corridor outside the diesel rooms. TVA evaluated the impact of these non-rated manway access openings in the FPR Part VII, Section 4.4, "Fire Barriers between DG [Diesel Generator] Storage Tank and DG Corridor." Section 6.2.9 of this evaluation provides the NRC staff evaluation of this deviation.

TVA evaluated the impact of an untested penetration assembly in the fire barrier between the fuel oil transfer pump room and the diesel generator corridor in the FPR Part VII, Section 4.6, "Fire Barriers between Fuel Oil Transfer Pump Room and Diesel Generator Building Corridor." Section 6.2.5 of this evaluation provides the NRC staff's evaluation of this deviation.

Based on its review of the information submitted by TVA, the NRC staff concludes that, with the exception of items evaluated elsewhere in this evaluation, the fire protection features for the diesel fuel oil storage areas are consistent with Position F.10 of Appendix A to BTP (APCSB) 9.5-1 and, therefore, are acceptable.

5.9   Safety-Related Pump Areas

5.9.1   CCS Pump Area

As described in TVA's response dated May 26, 2011 (ADAMS Accession No. ML111520119), to RAI FPR VII-3, the CCS pumps are located in the same fire area in the auxiliary building on elevation 713.0 feet. The two Train A CCS pumps are separated from the two Train B pumps, and the spare, by a partial height fire barrier.

TVA evaluated the partial height fire barrier between the CCS pumps and the ensuing redundant train separation issues in FPR Part VII, Section 2.5. Section 6.1.5 of this evaluation provides the NRC staff's evaluation of this configuration.

TVA stated in FPR Part VII, that the area containing the CCS pumps is provided with automatic pre-action sprinkler system protection at the ceiling and under the grated mezzanine over the CCS pumps as well as full coverage automatic smoke detection. Further, in FPR Part VI, TVA stated that hose stations are available to support manual fire-fighting.

Based on its review of the information submitted by TVA, the NRC staff concludes that, with the exception of items evaluated elsewhere in this evaluation, the fire protection features for the CCS pumps do not take any exceptions to Position F.11 of Appendix A to BTP (APCSB) 9.5-1 and, therefore, are acceptable.

5.9.2   Charging Pumps

As described in the FPR Part VI, each charging pump is located in its own 2-hour fire-rated compartment. TVA stated that the pump rooms and the corridor outside these rooms are protected by full coverage detection and an automatic pre-action sprinkler system. However, detection and suppression is not extended into the entrance labyrinth of the charging pump rooms. Further, TVA stated that hose stations are located in the corridor leading to these rooms and are available to support manual fire-fighting inside the pump rooms.

TVA evaluated the impact of the lack of total area suppression and detection in the FPR Part VII, Section 3.1, "Lack of Total Area Suppression and Detection." Section 6.1.7 of this evaluation provides the NRC staff evaluation of this deviation.

Based on its review of the information submitted by TVA, the NRC staff concludes that, with the exception of items evaluated elsewhere in this evaluation, the fire protection features for the charging pumps provide an equivalent level of safety to Position F.11 of Appendix A to BTP (APCSB) 9.5-1 and, therefore, are acceptable.

5.9.3   AFW Pumps

As described in the FPR Part VI, the two turbine-driven AFW pumps (one for each unit) are located in the auxiliary building on elevation 692.0 feet. Each pump is located in its own 2-hour fire rated fire compartment. TVA stated that each pump room is provided with full coverage automatic detection and an automatic pre-action sprinkler system. Further, TVA stated that hose stations are located in the corridor leading to these rooms and are available to support manual fire-fighting inside the pump rooms.

As described in TVA's response dated May 26, 2011 (ADAMS Accession No. ML111520119), to NRC question RAI FPR VII-3, the motor-driven AFW pumps (two per Unit) are located on opposite ends of the auxiliary building on elevation 713.0 feet. TVA stated that there is approximately 126 feet separating the Unit 1 and Unit 2 AFW pumps. TVA further stated that the area in which these pumps are located is protected by an automatic pre-action sprinkler system, and that automatic fire detection is provided throughout the area. TVA stated in FPR Part VI that hose stations are available in the area to support manual fire-fighting operations.

Based on its review of the information submitted by TVA, the NRC staff concludes that the fire protection features provided for the motor- and turbine-driven AFW pumps provide an equivalent level of fire safety to Position F.11 of Appendix A to BTP (APCSB) 9.5-1 and, therefore, are acceptable.

5.9.4   RHR Pumps

As described in the FPR Part VI, each RHR pump is located in its own 2-hour fire rated fire compartment. Each RHR pump room is a separate fire area and none of the rooms contain redundant trains of equipment or cables. TVA stated that the corridor outside these rooms has full coverage fire detection installed. In each pump room, fire detection is installed, except in the entrance labyrinths. TVA stated that the combustible loading in these rooms is insignificant, consisting mainly of the lube oil associated with the pump and valve. TVA stated that for each fire area, the capability to achieve safe shutdown has been demonstrated through analysis. Therefore, a fire in any of these fire areas will not endanger other safety related equipment required for safe plant shutdown. Further, TVA stated that hose stations are located in the corridor leading to these rooms and are available to support manual fire-fighting inside the individual RHR pump rooms.

Based on its review of the information submitted by TVA, the NRC staff concludes that the fire protection features for the RHR pumps provide an equivalent level of fire safety to Position F.11 of Appendix A to BTP (APCSB) 9.5-1 and, therefore, are acceptable.

5.9.5 ERCW Pumps

As described in FPR Part VI, the redundant ERCW pumps are separated by 3-hour fire rated barriers. These pumps are also separated from the traveling screen pumps by 3-hour barriers. However, these barriers have open scuppers at the base of the wall of the ERCW pump rooms.

TVA stated in FPR Part VI that heat detectors are installed over the ERCW pumps and that no redundant FSSD cables or equipment are installed in these areas. Further, TVA stated that manual fire suppression capability is available through use of hose stations installed in the ERCW strainer room and the screen wash pump room.

TVA evaluated the impact of the open scuppers in the fire barriers that separate the pumps from the traveling screens in the FPR Part VII, Section 2.6.2, "Justification for Scupper Openings." Section 6.2.7.2 of this evaluation provides the NRC staff evaluation of this configuration.

Based on its review of the information submitted by TVA, the NRC staff concludes that, with the exception of items evaluated elsewhere in this evaluation, the fire protection features for the ERCW pumps provide an equivalent level of fire safety to Position F.11 of Appendix A to BTP (APCSB) 9.5-1 and, therefore, are acceptable.

5.10 Other Plant Areas

5.10.1 Areas without Deviations or Evaluations

The NRC staff reviewed TVA's compliance with the following positions of Appendix A to BTP (APCSB) 9.5-1, as documented in FPR Part VIII:

- Position F.4 – "Plant Computer Room"
- Position F.6 – "Remote Safety-Related Panels"
- Position F.14 – "Radwaste Building"
- Position F.15 – "Decontamination Areas"
- Position F.16 – "Safety-Related Water Tanks"
- Position F.17 – "Cooling Towers"
- Position F.18 – "Miscellaneous Areas"

Based on its review of the information submitted by TVA, the NRC staff concludes that the fire protection features provided in these areas provides an equivalent level of fire safety as the guidance in these sections of Appendix A to BTP (APCSB) 9.5-1, and, therefore, are acceptable.

5.10.2 Areas with Deviations or Evaluations

The NRC staff reviewed TVA's compliance with the following positions of Appendix A to BTP (APCSB) 9.5-1, as documented in FPR Part VIII:

- Position F.12 – "New Fuel Area"
- Position F.13 – "Spent Fuel Pool Area"

TVA evaluated the impact of the lack of installed fire detection in these areas in FPR Part VII, Section 4.5. Section 6.2.1 of this evaluation provides the NRC staff's evaluation of this configuration.

Based on its review of the information submitted by TVA, the NRC staff concludes that, with the exception of items evaluated elsewhere in this evaluation, the fire protection features for these areas provide an equivalent level of fire safety to Positions F.12 and F.13 of Appendix A to BTP (APCSB) 9.5-1 and, therefore, are acceptable.

5.11   Specific Hazards

5.11.1 Hydrogen Piping

TVA stated in the FPR that a 1-inch seismically-designed hydrogen line is routed through the auxiliary building on elevation 713.0 feet to each unit's VCT. Two isolation valves are installed in the hydrogen supply line outside the auxiliary building. These valves close automatically when the downstream flow rate reaches 50 standard cubic feet per minute (scfm). TVA stated that any hydrogen leakage less than 50 scfm will be diffused and carried away by the auxiliary building ventilation system, keeping the hydrogen concentration in any given area below the lower explosive limit.

Based on its review of the information submitted by TVA, the NRC staff concludes that the hydrogen supply piping in the auxiliary building does not take any exceptions to Position D.2.b of Appendix A to BTP (APCSB) 9.5-1 and, therefore, is acceptable.

5.11.2 Transformers Installed Inside Buildings

TVA stated that transformers located inside of buildings are either dry type or medium voltage transformers that contain "high fire point," transformer liquid. The use of dry type transformers is consistent with the NRC guidance in Appendix A of BTP 9.5-1, Element D.1.g, but the use of transformers with "high fire point" silicone fluid is not included as part of the guidance.

In TVA's response dated August 5, 2011 (ADAMS Accession No. ML11224A052), to RAI VIII-21, TVA provided its justification for the use of the "high fire point" silicone fluid in lieu of the non-combustible liquid described in Appendix A to BTP 9.5-1. TVA stated that the non-combustible transformer liquids contained PCB fluids. PCB fluids are considered non-combustible, but constitute an occupational health and safety, as well as environmental, concern if leaked or spilled. Therefore, TVA decided to remove PCB fluids from the plant. Although the "high fire point" liquid is considered combustible, it is not considered flammable in accordance with the definition of flammable and combustible provided by NFPA 30-1973, "Flammable and Combustible Liquids Code."

TVA stated that all areas where these transformers are located have sprinkler protection. Based on the vendor information provided by TVA in Attachment 4 of its letter dated September 30, 2011 (ADAMS Accession No. ML13060A225), sprinkler systems are effective at extinguishing silicone fluid fires. TVA also considered dikes to contain the volume of the silicone fluid if it were to leak from the transformers.

In its response dated September 30, 2011, to RAI VIII-21.1, TVA provided additional information regarding the installation of transformers containing "high fire point" silicone fluid. The NRC staff questioned the location of these transformers in plant areas that constitute buffer zones between analysis volumes, since the transformers were not described as being located in the buffer zones. TVA confirmed, in its RAI response, that the transformers are not located in buffer zones for large fire areas except for in the electrical equipment room in the IPS.

The transformers in the electrical equipment room in the IPS have dikes, are protected with automatic fire suppression systems, and there is 20 feet of separation between the transformers and the redundant FSSD train. TVA stated that the 20 feet of separation has intervening combustibles, but the combustibles are not continuous. Therefore, in the event that a transformer fire was to occur in this area, automatic suppression and spatial separation is available to assure that safe shutdown capability is assured.

Based on its review of the information submitted by TVA, the NRC staff concludes that TVA's use of dry type transformers in plant areas is consistent with Appendix A of BTP 9.5-1, Element D.1.g, and, therefore, is acceptable. The use of "high fire point" silicone fluid in transformers in plant areas is acceptable where the transformers are installed in areas with automatic sprinkler systems and spatial separation, either buffer zones or 20 feet without continuous intervening combustibles, and where transformers have dikes large enough to contain the volume of the transformer fluid.

6.0   DEVIATIONS AND EVALUATIONS

In FPR Part VII, TVA documented deviations from WBN commitments against applicable NRC regulatory criteria and guidance documents and presented engineering evaluations of the adequacy of specific fire protection features.

6.1   Deviations and Evaluations Related to Criteria in Appendix R to 10 CFR Part 50

6.1.1   Deviation – Required Instrumentation for Alternative Shutdown

TVA committed to maintain safe shutdown capability during and after a fire in accordance with Section III.L of Appendix R to 10 CFR Part 50. Section III.L.2.d of Appendix R to 10 CFR Part 50 states that the process monitoring function for alternative shutdown be capable of providing direct readings of the process variables necessary to perform and control a plant cooldown.

Contrary to Section III.L.2.d of Appendix R to 10 CFR Part 50, TVA has not provided instrumentation in the ACR for (1) tank level indication for the CST or the RWST, (2) wide-range SG level indication, and (3) cold-leg temperature indication. TVA evaluated these deviations in FPR Part VII, Section 2.1. TVA's justification for omitting this instrumentation is given below.

The CST level indication is not considered essential in the ACR because automatic switchover of the AFW pump suction from the CST to the ERCW header is independent of the control building, and therefore would be available when control is established in the ACR.

The RWST level indication is not considered essential in the ACR because the RWST contains almost 20 times the inventory required for cold shutdown. Because the RWST is primarily used as makeup water for contraction resulting from cooldown over a period of hours, the excess inventory in the RWST is considered sufficient without level indication in the ACR.

Wide-range SG level indication is not provided in the ACR. Instead, the narrow-range SG level and AFW flow indication to each SG are provided in the ACR and are sufficient for use in safe shutdown procedures whenever the ACR is utilized. This instrumentation also provides input to the automatic control utilized to maintain SG level during plant shutdown from the ACR. Although wide-range instrumentation is available in the MCR, no automatic control or safety system inputs are derived from this instrumentation. Therefore, the AFW flow indication is sufficient for the operator to confirm that adequate post-trip SG inventory is available in the event that SG level falls below the range of the narrow range indicators that are located in the ACR.

Cold leg temperature indication is not provided in the ACR. Cold leg temperature (T_C), is used for monitoring natural circulation. Rather than using T_C, TVA monitors natural circulation by inferring T_{SAT}, the saturation temperature corresponding to the secondary-side SG pressure. In the natural circulation mode of operation, the difference between the hot-leg and cold-leg temperature (T_H-T_C) provides an effective indication of when natural circulation is established and whether it is being maintained. T_{SAT} will be used to monitor natural circulation in the reactor coolant loop in the operating range from full power to the hot standby condition. To demonstrate that T_{SAT} will accurately monitor natural circulation in the operating range from hot standby to cold shutdown, TVA analyzed the correlation between T_{SAT} and T_C while a reactor was brought to cold-shutdown condition.

TVA stated that the Westinghouse Owners Group document "Emergency Response Guidelines, Generic Issue on Natural Circulation," Revision 1, provides specific guidelines on how an operator can verify that natural circulation has been established without T_C being available. The Westinghouse Owners Group recommends the use of the following criteria for verifying natural circulation: (1) RCS is subcooling (conversion of pressurizer pressure to T_{SAT} and subtracting T_H); (2) T_H is stable or decreasing, and (3) SG pressure is stable or decreasing. The instrumentation needed to use these methods of verifying natural circulation is available to the operator in the ACR. Therefore, the installed indication is sufficient to compensate for the lack of T_C indication in the ACR.

Based on its review of the information submitted by TVA, the NRC staff concludes that not providing wide-range SG level, CST and RWST tank water level indication, and cold-leg temperature indication in the ACR, are acceptable deviations from Section III.L.2.d of Appendix R to 10 CFR Part 50.

6.1.2   Deviation – Noncombustible Radiant Energy Heat Shields

TVA committed to maintain safe shutdown capability during and after a fire in accordance with Section III.G of Appendix R to 10 CFR Part 50. Section III.G.2.f of Appendix R to 10 CFR Part 50 states that inside non-inerted containments, separation of cables and equipment and associated non-safety circuits of redundant trains by a noncombustible RES is an acceptable method of ensuring that a redundant train of equipment and circuits are protected from a fire.

The acceptance criteria included in previous revisions to NUREG-0800, "Standard Review Plan for the Review of Safety Analysis Reports for Nuclear Power Plants: LWR Edition," Chapter 9, "Auxiliary Systems," Section 9.5.1.1, "Fire Protection Program," as BTP APCSB 9.5-1 (BTP Chemical Engineering Branch (CMEB) 9.5-1 at the time of publication of GL 86-10) have been removed and have been incorporated in Revision 2 of RG 1.189. Section 6.1.1.1, "Containment Electrical Separation," to RG 1.189 states the following:

> Inside noninerted containments, one of the fire protection means specified in Regulatory Position 5.3.1.1, or one of the following, should be provided:
>
> a. separation of cables and equipment and associated nonsafety circuits of redundant trains by a horizontal distance of more than 6.1 m (20 feet) with no intervening combustibles or fire hazards,
>
> b. installation of fire detectors and an automatic fire suppression system in the fire area, or
>
> c. separation of cables and equipment and associated nonsafety circuits of redundant trains by a noncombustible RES having a minimum fire rating of 30 minutes, as demonstrated by testing or analysis.

Section 3.7.1, to GL 86-10 states the following:

> The guidelines in BTP CMEB 9.5-1, Section C.7.a.(1)b. indicate that these shields should have a fire rating of 1/2 hour. In our opinion any material with a 1/2 hour fire rating should be capable of performing the required function.

TVA evaluated this deviation in FPR Part VII, Section 2.2. The RESs installed inside the reactor buildings at WBN are Minnesota Mining and Manufacturing (3M) M20A in the annulus, and M20C in the Unit 1 primary containment. TVA stated that site calculations EPM-BFS-041895 and EPM-BFS-053195 provide the design basis for the number of layers of M20A and M20C required to provide approximately 1/2 hour RESs for electrical raceways containing circuits required for FSSD. These calculations were based on fire tests performed by 3M to UL Subject 1724, "Fire Tests for Electrical Circuit Protective Systems." The fire exposure used in the tests is the standard time-temperature curve from ASTM E119.

TVA had a series of fire resistance tests performed on the material at Omega Point Laboratories for combustibility of the installed materials. The 3M M20A and M20C did not meet the criteria for non-combustibility per ASTM E136, "Standard Test Method for Behavior of Materials in a Vertical Tube Furnace at 750° C." Additional fire tests to the criteria in ASTM E1354, "Standard Test Method for Heat and Visible Smoke Release Rates for Materials and Products Using an Oxygen Consumption Calorimeter," were performed with various RES materials. The results indicated that the peak heat release rate (HRR) and the total heat release rate (THR) for the 3M M20A and M20C was lower than that of marinite board. Since marinite board is accepted in GL 86-10 as an acceptable RES material, and the 3M materials used at WBN have lower HRR and THR than marinite board, the 3M materials are also considered sufficiently noncombustible for the use as RES.

Based on its review of the information submitted by TVA, the NRC staff concludes that the fire rating and combustibility of the 3M M20A in the annulus and 3M M20C in the WBN Unit 1 primary containment provide an equivalent level of fire safety to that required by Section III.G.2.f of Appendix R and, therefore, are acceptable.

6.1.3 Deviation – Lack of Automatic Fire Suppression in Alternative Shutdown Locations

TVA committed to maintain safe shutdown capability during and after a fire in accordance with Section III.G of Appendix R to 10 CFR Part 50. Section III.G.3 of Appendix R to 10 CFR Part 50 states that fire detection and a fixed fire suppression system shall be installed in the areas, rooms, or zones requiring alternative or dedicated shutdown capability.

TVA requested a deviation from this Appendix R requirement for a number of control building rooms that lack fixed fire suppression, and some rooms that also lack fire detection.

The control building is separated from the ACR and adjacent plant areas by equivalent 3-hour fire rated barriers except for the equipment hatch in the ceiling separating the control building from the turbine building. The justification for the hatch opening through the ceilings of Rooms 692.0-C1 and 692.0-C10 to the turbine building is evaluated in Section 6.2.7.4 of this evaluation. The turbine building is separated from the ACR and adjacent plant areas by equivalent 3-hour fire rated barriers. This separation provides assurance that safe shutdown capability is assured for a fire in the control building.

All the control building rooms that lack fixed fire suppression have limited ignition sources and low or insignificant combustible loading. In addition, all of the rooms have standpipes and hose stations available for manual fire-fighting. Only a few rooms lack full area detection. These rooms are stairwells, shower rooms, the telephone room, and the space above the living area on the 755.0 foot elevation. Frequent use of the stairwells would lead to discovery of a fire in its early stages and would also reduce the likelihood that combustibles could accumulate there. The other rooms all are described as having negligible combustible loading.

Based on its review of the information submitted by TVA, the NRC staff concludes that the lack of fire detection and fixed suppression in the control building areas identified above is an acceptable deviation from the requirements of Section III.G.3 of Appendix R to 10 CFR Part 50, because all rooms that lack fixed suppression have low levels of combustibles and available manual suppression, and the rooms that also do not have fire detection have negligible fire loading.

6.1.4 Deviation – Intervening Combustibles

TVA committed to maintain safe shutdown capability during and after a fire in accordance with Section III.G of Appendix R to 10 CFR Part 50. Section III.G.2.b of Appendix R to 10 CFR Part 50 states that separation of redundant trains of safe-shutdown cables and equipment by a horizontal distance of more than 20 feet with no intervening combustibles. In addition, fire detection and an automatic fire suppression system shall be installed in the area.

In FPR Part VII, Section 2.4, TVA requested a deviation from compliance with Section III.G of Appendix R to 10 CFR Part 50 for 20 feet horizontal distance with no intervening combustibles for safe shutdown components and cables in the auxiliary building and the IPS electrical equipment room. WBN stated that safe shutdown components in the auxiliary building and IPS electric equipment room are in compliance with Section III.G.2.b of Appendix R to 10 CFR Part 50 requirements except that intervening combustibles are located between the redundant components.

The intervening combustibles in the auxiliary building are mainly in the form of insulation on cables in open ladder type cable trays and Thermo-Lag fire barrier material. The remaining in situ combustible loading consists of lubricating oil in pumps, motors, and valves; transformer silicon liquid; and plastics in electrical panels, junction boxes, etc. The intervening combustibles in the IPS electric equipment room are mainly in the form of insulation on cables in open ladder type cable trays and transformer silicone liquid. The remaining in situ combustible loading consists of lubricating oil in small pumps, plastics associated with electrical panels, junction boxes, etc. Discussion of the nature of the transformer silicon liquid can be found in Section 5.11.2 of this evaluation.

The presence of these intervening combustibles is a concern because they add to a fire's intensity at the ceiling and they could serve as a path for fire propagation between the redundant safe-shutdown trains.

For intervening combustibles in the auxiliary building, TVA stated that existing sprinkler heads, which are capable of fully developing spray patterns at the ceiling, provide acceptable floor coverage if there are no intermediate obstructions in their patterns, which are greater than 48 inches wide. Additional intermediate sprinklers are provided for 48 inch wide obstructions and for combinations of obstructions that, when overlapped, constitute a 48 inch wide obstruction, that overlap or combinations of obstructions have less than a 4 inch flue space between them when viewed from immediately below. No combination of obstructions may traverse the 4 inch flue space and block more than 2 feet of any 8 feet of flue space. To mitigate the effects of an exposure fire from transient combustibles at the floor level, TVA stated that floor level sprinkler coverage is provided under intermediate obstructions for up to a 30 foot wide path where spatially separated redundant FSSD components exist.

TVA stated that for intervening combustibles in the IPS electrical equipment room, sprinkler protection has been provided at the ceiling level. Due to the presence of obstructions such as HVAC ducts, cable trays, pipes, and supports, these systems have been upgraded. Sprinkler heads were added to provide full coverage at the ceiling level and to compensate for large intermediate level obstructions. To mitigate the effects of an exposure fire from transient combustibles at the floor level, TVA provided floor level sprinkler coverage under intermediate obstructions for up to a 30-foot wide path for spatially separated redundant FSSD components.

TVA concluded that, if a fire were to occur, these sprinkler systems would develop effective spray patterns at the ceiling, and the water would cascade down through the cable trays in the intervening spaces. The cooling effect of these sprinklers, once actuated, would help cool the layer of hot gas at the ceiling, prevent the formation of a high temperature plume, and cool the room. The sprinklers under the intermediate level obstructions would actuate to ensure that floor level coverage is provided under the obstructions. In addition, the coverage provided by the ceiling sprinklers would produce sufficient cooling to reduce the likelihood that fire will propagate across the intervening space between the redundant trains.

Based on its review of the information submitted by TVA, the NRC staff concludes that, because of the arrangement and nature of the combustibles, and the upgraded suppression systems, the presence of intervening combustibles as fire hazards between redundant trains of safe shutdown functions is an acceptable deviation from the requirements of Section III.G.2.b of Appendix R to 10 CFR Part 50.

6.1.5   Deviation – Partial Fire Wall between CCS Pumps

TVA committed to maintain safe shutdown capability during and after a fire in accordance with Section III.G of Appendix R to 10 CFR Part 50. Section III.G.2.b of Appendix R to 10 CFR Part 50 states that separation of cables and equipment and associated non-safety circuits of redundant trains by a horizontal distance of more than 20 feet with no intervening combustibles or fire hazards. In addition, fire detectors and an automatic fire suppression system shall be installed in the fire area. Section III.G.2.c states that enclosure of cables and equipment and associated non-safety circuits of one redundant train in a fire barrier having a 1-hour rating. In addition, fire detectors and an automatic fire suppression system shall be installed in the fire area.

In FPR Part VII, Section 2.5, TVA requested a deviation from these Appendix R requirements for redundant CCS pumps that are protected by fire detectors and an automatic fire suppression system, but are separated by a partial height and width noncombustible wall.

The five CCS pumps are located in Fire Area 8, Room 713.0-A1, in subsections 713.0-A1A1, -A1A2 and -A1A3, on elevation 713.0 feet of the auxiliary building. The two Train B pumps are separated from both Train A pumps and the spare pump by a noncombustible wall which extends 3 feet above the highest point of the pumps. A ceiling-level pre-action sprinkler system is provided for cable tray and general area coverage. Automatic sprinkler coverage has also been provided under the pipe-break barrier for the Unit 1 motor-driven AFW pumps and under the mezzanine for all five CCS pumps. Cross-zoned ionization smoke detectors are provided to actuate the pre-action suppression systems and give early warning of a fire.

The combustibles in Room 713.0-A1 consist of lube oil in the pumps, motors, and valves; plastics associated with the electrical panels, boxes and lights, insulation on cables routed in cable trays; and anticipated amounts of radwaste trash and laundry. The fire severity for this room is classified as moderately severe. However, TVA stated that approximately 95 percent of the in situ combustible loading in this area is due to the insulation on cables routed in cable trays and the Thermo-Lag fire barrier material. The majority of the remaining combustible loading in the immediate area of the CCS pumps is due to the approximately 6 gallons of lube oil associated with each CCS pump and approximately 45 gallons of lube oil associated with each of the two Unit 1 AFW pumps. The cables are protected electrically with appropriately sized circuit protective devices (breakers and fuses) that will actuate on electrical faults prior to the jacket material of faulted cables reaching their auto-ignition temperature. A fire due to transient combustibles located near the edge of the partial height fire barriers would not pose a threat to more than one CCS pump due to the lack of combustibles. Additionally, raceways containing the redundant cables for the CCS pumps are separated by 20 feet or more or by noncombustible barriers.

Based on its review of the information submitted by TVA, the NRC staff concludes that the fire detection system and automatic sprinkler system would detect and suppress a fire prior to becoming a threat to the redundant pumps on the other side of the noncombustible barrier. Until the fire is suppressed, the noncombustible barrier will shield the pumps from radiant heat on one side and from fire on the other. Therefore, because of the noncombustible nature of the barrier, the installed fire detectors and automatic fire suppression systems, and redundant cable separation, the partial height fire wall is an acceptable deviation from the technical requirements of Sections III.G.2.b of Appendix R to 10 CFR Part 50.

6.1.6   Deviation – Emergency Lighting

TVA committed to provide emergency lighting to assure safe shutdown capability is maintained during and after a fire in accordance with Section III.J of Appendix R to 10 CFR Part 50. Section III.J of Appendix R to 10 CFR Part 50 states that emergency lighting units with at least an 8 hour battery power supply be provided in all areas needed for operation of safe-shutdown equipment and for necessary access and egress routes.

In FPR Part VII, Section 2.7, TVA requested a deviation from this emergency lighting requirement in each containment, the turbine building, and the yard. Dedicated and maintained hand-held portable lanterns are provided in lieu of installed battery pack lighting units in both containments. Emergency diesel generator backed standby lighting is installed and maintained for the turbine building. Security diesel generator backed standby lighting is installed and maintained for the yard. Additionally, hand-held portable lanterns are available to supplement yard and turbine building diesel backed lighting systems to provide additional task lighting capability.

The hand-held portable lanterns used for this purpose are rechargeable, industrial duty, 12 VDC devices. They can continuously operate for up to 9 hours per charge. They are stored in cages and placed on electrical charge from the plant's 120 VAC lighting system. The lights are inventoried to ensure they are in their assigned location, operated to ensure they will illuminate, and are verified to be on charge every 13 weeks.

Based on its review of the information submitted by TVA, the NRC staff concludes that the use of installed standby lighting and hand-held portable lighting units for the yard and turbine building is an acceptable deviation from the lighting criteria required by Section III.J, of Appendix R to 10 CFR Part 50, and, therefore, is acceptable.

OMAs requiring entry into primary containment would only result from fire damage to the RHR isolation valves or cables near the valves which are located inside lower containment. The OMAs to align the RHR isolation valves may be performed anytime within 4 hours after reactor trip. This allows ample time to extinguish the fire, obtain the portable lanterns, and operate the valves. As described above, WBN has dedicated hand-held portable lighting units for use in supporting manual fire-fighting and safe shutdown OMAs for fires in the lower containment.

A fire affecting the RHR isolation valves could damage lighting circuits in the immediate vicinity, but would not be expected to disable all lower containment lighting, since different circuits are used at each elevation. Additionally, two standby lighting circuits, with fixtures strategically located throughout lower containment, provide lighting in case of fire damage to the normal lighting cabinet.

TVA's concerns regarding the installation of 8-hour emergency lighting units inside containment include the reduced life of the batteries in the high temperature and humidity environment experienced inside the primary containment. Also, ALARA concerns would limit testing and maintenance to reactor outages, since access into the primary containment during plant operations is restricted.

Based on its review of the information submitted by TVA, the NRC staff concludes that, based on the complications of testing and maintaining 8-hour fixed emergency lighting units, and TVA's design description of the installed lighting in the lower containment complemented by the dedicated hand-held portable lighting units, the installation of 8-hour emergency lighting units is unnecessary to provide access and egress to the manual action sites and perform safe shutdown actions in primary containment. Therefore, the use of installed lighting and hand-held portable lighting units for this area is an acceptable deviation from the lighting criteria required by Section III.J of Appendix R to 10 CFR Part 50.

6.1.7   Evaluation – Lack of Total Area Suppression and Detection

TVA committed to meet Section III.G.2 of Appendix R to 10 CFR Part 50 for hot shutdown capability, which states that when redundant trains of cables or equipment necessary for post-FSSD are installed in the same fire area, fire detectors and automatic fire suppression must be installed, unless one train is protected by a 3-hour rated fire barrier. Position 5 of the Attachment to GL 86-10 states that to meet the requirements of Section III.G.2 of Appendix R to 10 CFR Part 50, less than full area coverage may be adequate to comply with the regulation if the suppression and detection installed is sufficient to protect against the hazards of the fire area.

In FPR Part VII, Section 3.1, TVA evaluated portions of fire areas that contain both trains of safe shutdown success paths, but do not have full coverage fire detection and suppression installed. The WBN plant has some fire areas that include multiple subdivisions, called rooms.

These rooms may not be separated from the other rooms within the fire area by rated fire barriers.

The NRC staff notes that for fire areas composed of multiple rooms, the rooms which contain redundant safe shutdown equipment have either 3-hour rated barriers to protect one train of the safe shutdown equipment, or the rooms are equipped with fire detection and automatic suppression, and have some spatial separation between trains (see Section 3.2.1 of this evaluation). Therefore, these rooms are not considered to be credible exposure hazards to the other rooms in the fire area that have redundant safe shutdown equipment.

Some of the rooms contain safe shutdown equipment, but there is not redundant safe shutdown equipment required for hot shutdown in the room. In other cases, the safe shutdown equipment is needed for cold shutdown, or for alternative shutdown. In still other cases, the safe shutdown equipment is not used to provide for plant safe shutdown for a fire in the room; that is, it is relied upon for a fire elsewhere in the plant. In any of these cases, safe shutdown equipment is available outside of the room if there is a fire in the room and any exposure hazard in the room to another room would be mitigated by the protection in the other room.

Based on the information provided by TVA, there are rooms that lack full area fire detection and suppression that do not contain redundant safe shutdown equipment needed for hot shutdown and do not constitute exposure hazards to other rooms within the fire area. The NRC staff has reviewed this information and concludes that this is acceptable.

The descriptions in the evaluations state that the plant provided only one train of FSSD equipment and cables in Centrifugal Charging Pump (CCP) Rooms 1B-B (Room 692.0-A10; Fire Area 6), 2A-A (Room 692.0-A22; Fire Area 67), and 2B-B (Room 692.0-A23; Fire Area 68). However, Fire Areas 6, 67, and 68 consist solely of the single CCP room. Because these rooms do not contain redundant trains of equipment or cables, the NRC staff did not review these evaluations.

Rooms that contain redundant cables or equipment necessary for post-FSSD

480 V Board Rooms 1B (Room 772.0-A2; Fire Area 33) and 2B (Room 772.0-A15; Fire Area 45)

In FPR Part VII Section 3.1.8, TVA stated that in 480 V Board Rooms 1B (Room 772.0-A2; Fire Area 33) and 2B (Room 772.0-A15; Fire Area 45), pre-action sprinkler systems are provided throughout both rooms except for the portion of each room that contains one set of vital battery inverters and chargers. Additionally, ionization detection is installed throughout both rooms. TVA further stated that the redundant inverters and chargers and associated cables are separated by a minimum of 42 feet and are located at opposite ends of each room. Additionally, TVA stated that other redundant components in the rooms are located within the suppressed area of each room and are separated in accordance with Section III.G.2 of Appendix R to 10 CFR Part 50. A fire in the unsprinklered locations in these rooms would be detected by the installed fire detection systems before propagating significantly. If the fire propagated rapidly before the fire brigade arrived, individual sprinklers in the protected portions of the rooms would operate to limit the spread of fire and to protect the redundant systems until the fire was controlled and suppressed by the plant fire brigade.

Based on its review of the information submitted by TVA, the NRC staff concludes that the partial coverage of the automatic suppression systems in these rooms is sufficient to protect against the fire hazards in these areas and that this level of protection, including the separation between trains, provides an equivalent level of fire safety to that required by Sections III.G.2.b and III.G.2.c of Appendix R to 10 CFR Part 50 and, therefore, is acceptable.

6.1.8   Evaluation – Reactor Coolant Pump Oil Collection System

TVA has committed to meet Section III.O of Appendix R to 10 CFR Part 50. This section states, in part, that RCPs be equipped with an oil collection system if the containment is not inerted during normal operation and that the system be capable of collecting lube oil from all potential pressurized and unpressurized leakage sites in the RCP lube oil system.

In FPR Part VII, Section 2.8, TVA stated that the RCP oil collection system must function in an area with significant ventilation airflows from both the control rod drive mechanism cooling units and the RCP motor itself. A minor leak in the lubrication system that causes oil to drip in an area where the ventilation airflow is strong can result in the oil becoming entrained in ventilation air, which in turn could prevent the leak from ever entering the collection system. The need for ventilation around the RCP dictates that some ventilation flow areas must be present in areas around the lube oil system and the oil collection system. In designing the oil collection system, it is not feasible in all instances to prevent minor amounts of oil from becoming entrained in the ventilation air and escaping the collection system. This oil may become a thin film on the piping mirror insulation and supports in the vicinity of the RCPs.

TVA described the RCP oil collection systems in a letter dated May 26, 1995 (ADAMS Accession No. ML073230888). TVA used the following design criteria as the basis for the oil collection systems.

The oil collection system on each RCP collects oil from all potential leakage locations, including the RCP oil lift pump, system piping, overflow lines, the lube oil cooler, oil fill and drain lines, flanged connections on the oil lines, and the lube oil reservoirs. Each RCP oil collection system consists of spray shields/deflectors, a collection basin, a lift pump collection tray, a lower bearing collection tray and drain, drain piping, and a closed, vented container (reactor building floor and equipment drain sump).

The drain piping from each RCP's oil collection basin is directed to a drain header. The drain header runs through the shield wall and into the raceway area inside primary containment and runs through the floor into the 1600 gallon capacity sump. As required by Appendix R, the sump is a closed container and is equipped with a flame arrester on the vent line. Each unit's sump has sufficient capacity to hold the entire RCP oil inventory of all four RCPs.

TVA stated that up to 14 gallons of oil could collect in the lower motor support housing before beginning to drain to the collection system. The RCPs are equipped with control loop level indication that would initiate an alarm in the MCR if 2 or more gallons of lube oil are lost from the RCP. Collection of oil within the lower motor support housing is acceptable since the oil, and possible fire, would be contained within the RCP and would not impact surrounding equipment such that safe shutdown could be affected. In addition, the RCP is equipped with a water-based fire suppression system such that a fire at the RCP would have automatic suppression available.

The RCP pumps, lubricating oil systems, oil spray shields, oil collection basins, drain piping, and containment sumps are designed to seismic Category I requirements so as not to fail during a safe-shutdown earthquake.

Each of the four RCPs is protected by a fixed fire suppression and detection system. A heat collection hood is installed directly above the RCP motors. Each of the RCPs is protected by a separate closed-head pre-action automatic water spray system that is installed under this hood. Each system has a ring header containing eight nozzles. The header is located approximately 4 feet above the top of the RCP motor and the nozzles, which actuate at 500 °F, are oriented so as to provide optimum coverage of the RCP motor from above. In addition, there are four rate-compensating/fixed-temperature spot-type thermal detectors located above the RCP motors on the bottom side of the heat-collection hood. These detectors are Class A supervised, have a thermal rating of between 200 °F and 225 °F and are alarmed and annunciated in the MCR. In the event of a fire, this hood acts as a ceiling, forcing the heat to stall around the detectors and the suppression nozzles, thus reducing the response time of these fire protection devices.

Based on its review of the information submitted by TVA, the NRC staff concludes that the RCP oil collection systems have been designed in accordance with Section III.O of Appendix R to 10 CFR Part 50. The deviations to allow collection of oil in the lower motor support housing and to allow minor amounts of oil to escape the oil collection system and become a thin film on piping mirror insulation and supports in the vicinity of the RCPs, are acceptable since large leakages would be alarmed to the control room and the RCP cubicles are equipped with fixed fire suppression and detection is provided.

6.1.9   Evaluation - Unit 2 Manual Actions

TVA committed to meet Section III.G of Appendix R to 10 CFR Part 50. Section III.G of Appendix R to 10 CFR Part 50 provides a number of acceptable methods of providing reasonable assurance that one of the safe shutdown trains is free of fire damage using a combination of physical separation, fire wraps, fire detection and fire suppression. Unless previously approved by the NRC, the use of OMAs is not a means of assuring that a safe shutdown train is free of fire damage, as described in Section III.G of Appendix R to 10 CFR Part 50. Discussion of OMAs needed for equipment important for safe shutdown is included in Section 3.5 of this evaluation.

TVA developed evaluations to demonstrate that OMAs are capable of accomplishing various safe shutdown functions and terminating spurious equipment operations that have the potential to interfere with safe shutdown. TVA also described the fire protection defense-in-depth features within each room that reduces the likelihood that an OMA would be needed. The NRC staff has reviewed the OMAs in the below captioned rooms.

Operator Manual Action Number	Room of Postulated Fire
OMA-1016	713.0-A1B‡, 737.0-A1B, 737.0-A1N*, 757.0-A1, 757.0-A5, 757.0-A10, 757.0-A17, 757.0-A22, 757.0-A24, 757.0-A28, 772.0-A2 East, 772.0-A5, 772.0-A8, 772.05-A15* East, 772.0-A15 West*, 782.0-A1, 782.0-A2

Operator Manual Action Number	Room of Postulated Fire
OMA-1022	713.0-A1A‡, 713.0-A27, 729.0-A8, 737.0-A1A, 737.0-A5S, 737.0-A9M, 737.0-A9N, 737.0-A9S, 757.0-A2, 757.0-A4‡, 757.0-A9, 757.0-A16, 757.0-A23, 757.0-A27, 772.0-A1, 772.0-A2 West, 772.0-A4‡, 772.0-A8, 772.0-A9, 772.0-A10, 772.0-A16, 782.0-A3, 782.0-A4
OMA-1023	713.0-A1A‡, 713.0-A1B‡, 713.0-A27, 729.0-A8, 737.0-A1A, 737.0-A1C, 737.0-A5M, 737.0-A5N, 737.0-A5S, 737.0-A9M, 737.0-A9N, 737.0-A9S, 757.0-A2, 757.0-A4‡, 757.0-A9, 757.0-A16, 757.0-A21, 757.0-A23, 757.0-A27, 772.0-A1, 772.0-A2 West, 772.0-A4‡, 772.0-A6, 772.0-A8, 772.0-A9, 772.0-A10, 772.0-A12, 772.0-A16, 782.0-A3, 782.0-A4, DBIPS-A‡, IPS-A‡, IPS-C Middle, IPS-C West
OMA-1024	737.0-A1B, 737.0-A1N*, 737.0-A12, 757.0-A1, 757.0-A3‡, 757.0-A5, 757.0-A10, 757.0-A17, 757.0-A22, 757.0-A24, 757.0-A26, 757.0-A28, 772.0-A2 East, 772.0-A8, 772.0-A11, 772.0-A15* East, 772.0-A15 West*, 782.0-A1, 782.0-A2, DBIPS-B‡, IPS-B‡, IPS-C East
OMA-1065	692.0-A25
OMA-1066	692.0-A25†
OMA-1159 & 1160	692.0-A1B*, 692.0-A22*
OMA-1275	713.0-A1B‡
OMA-1444 & 1445	772.0-A15 East*‡, 772.0-A15 West*‡, 772.0-A16‡
OMA-1448	772.0-A15 West*‡
OMA-1488	772.0-A13‡
OMA-1489	772.0-A14‡
OMA 1495 & 1496‡	772.0-A15 West*
OMA-1515	713.0-A1B‡
OMA-1516 & 1517	757.0-A21‡
OMA-1535 & 1536	737.0-A1N*‡
OMA-1540 & 1542	737.0-A1B‡

Key:
* Lacks full area fire detection, automatic suppression, or other defense-in-depth features – reviewed as part of separate deviation.
† OMA involves re-entry into room with postulated fire.
‡ OMA for this area either lacks full detection, full suppression, or 40 minutes of time margin, or a combination of these features.

TVA used the guidance in NUREG-1852, "Demonstrating the Feasibility and Reliability of Operator Manual Actions in Response to Fire," in determining the feasibility and reliability of manual actions. The following criteria were used to consider feasibility and reliability: (1) Adequate Time Available to Perform Actions; and (2) Adequate Time Available to Ensure Reliability. Most manual actions that had at least 40 minutes of margin were considered feasible and reliable. This considered the estimated travel and performance of the OMA, based on time trials performed for Unit 1 OMAs. The diagnosis time for OMAs is discussed in Section 3.5.4 of this evaluation. For the OMAs listed above, demonstrations have been performed to show that there will be at least 40 minutes of remaining margin upon the completion of the OMAs.

Some of the OMAs that had less than 40 minutes of margin have been designated in the table above as having reduced feasibility and reliability. Each of these OMAs have been evaluated specifically below with due consideration to time margin, defense-in-depth features, and other characteristics that would impact the likelihood that the manual action would be needed and the likely successful performance of the manual action.

In some cases, re-entry of the fire area was considered after 60 minutes. These areas have automatic fire suppression, therefore the NRC staff considers re-entry in these cases to be acceptable. In other cases, the OMA may be needed in a room, but not the same room, on the same elevation as the postulated fire. In this event, TVA considered the possible environmental effects of the fire and concluded that those factors would not prevent the performance of the OMA. These areas had full suppression and detection, therefore this is considered acceptable by the NRC staff.

TVA also considered: (1) Environmental Factors, (2) Equipment Functionality and Accessibility, (3) Available Indications, (4) Communications, (5) Portable Equipment, (6) Personnel Protection Equipment, and (7) Procedures and Training. TVA stated that the above criteria do not adversely affect the performance of the action for the Unit 2 OMAs.

Fire protection defense-in-depth features, such as fire prevention, fire detection, and fire suppression apply to each of these rooms. Most rooms have full area fire detection and automatic fire suppression. For those areas that lack full fire area detection and automatic fire suppression, they are designated in the table with an asterisk. TVA performed an analysis of the fire hazards in the area and determined that the fire hazards in the area do not warrant the installation of fire detection and automatic suppression. These areas are typically pipe chases, tunnels, tank rooms, labyrinth entrances, corridors, or portions of larger rooms where the majority of the room is protected. The review of these systems is described in Section 6.1.7 of this evaluation.

Some rooms lack the typical full area detection and suppression and have not been evaluated previously. The FPR includes a description of the protection features in the room. For those rooms that lack full fire detection and automatic fire suppression, the NRC staff has evaluated these areas specifically below. This evaluation uses the available defense-in-depth information and information about the OMA to determine if having less than full area suppression and detection is acceptable. Using full area suppression and automatic detection as criteria for OMAs is not intended to imply that they are required; rather, the NRC staff deemed full detection and automatic suppression as a robust level of protection. Less than full detection and automatic suppression received a more detailed review by the NRC staff.

In its response by letter dated September 30, 2011 (ADAMS Accession No. ML13060A225), to NRC question RAI FPR VII-24, TVA stated that the operators that will be performing the manual actions could be working anywhere in the plant and would be summoned to the MCR, or the ACR for a control building fire, upon the confirmation of a fire. Upon arrival at the control room, the operators will receive their assignments and procedures. TVA included a description of where these OMAs are performed in sequence with other OMAs. Based on the information provided by TVA, the NRC staff concludes that the number of available operators is sufficient to perform the manual actions.

Evaluation of OMAs needed for fires in areas that lack fire detection or automatic suppression or both

Rooms 713.0-A1A and 713.0-A1B lack automatic suppression above the boric acid tanks. TVA stated that the area of the boric acid tanks is considered a combustible control zone, and that the automatic suppression in the immediate area would control the spread of fire. Therefore the NRC staff concludes this lack of full area suppression is acceptable.

Rooms 757.0-A3, 757.0-A4, 757.0-A22, 757.0-A23, 772.0-A4, 772.0-A13, and 772.0-A14, are equipped with ionization fire detection systems and manual fire sprinkler systems, rather than the typical automatic fire sprinkler system. These vital battery board rooms are described as having battery and instrument boards, transformers, control panels, and junction boxes. These combustibles are considered small as compared to typical power plant electrical panels, and TVA does not consider them credible ignition sources due to proper circuit protection and low concentrations of combustibles. Transient combustibles are controlled by plant procedures. OMAs 1016 (757.0-A22), 1022 (757.0-A4, 757.0-A23, and 772.0-A4), 1023 (757.0-A4, 757.0-A23, and 772.0-A4), 1024 (757.0-A3 and 757.0-A22), 1488 (772.0-A13), and 1489 (772.0-A14) each have at least 40 minutes of time margin. Based on the installed ionization fire detection systems, the manual fire suppression systems, the limited combustibles, and the available time margin, the NRC staff finds this evaluation is acceptable.

Rooms DBIPS-A and DBIPS-B, the IPS Duct Banks, have no detection or suppression. These areas have no credible ignition sources for the installed cables in the area. Since these are underground electrical conduit banks, no transient combustibles are expected. The manual actions that may be needed for fires in these duct banks are OMAs 1023 (DBIPS-A) and 1024 (DBIPS-B), and each have a 40-minute time margin. Based on the limited ignition sources for this underground duct bank, and the available time margin, the NRC staff finds these evaluations are acceptable.

Rooms IPS-A and IPS-B, the IPS areas A and B, have fire detection over the ECRW pumps and in each of the ERCW strainer rooms. Each area has a floor area in excess of 3500 square feet and a ceiling height of at least 13 feet. The combustibles in the room consist of the lubricating oil associated with the pumps, transformers, and MCCs. The OMAs 1023 (IPS-A), and 1024 (IPS-B) have 40-minute time margin. Based on the partial detection, the size of the rooms, and the available time margin, the NRC staff finds these evaluations are acceptable.

Evaluation of OMAs that lack 40 minutes of time margin

OMA 1275, for a fire in Room 713.0-A1B, lacks the typical minimum time margin of 40 minutes. The time margin for this action is analyzed to be 12 minutes for an action needed to be performed in 20 minutes. The demonstrated time for the comparable Unit 1 action was less than 8 minutes. This is the first action that the operator performing the OMA will do based on the analysis. The fire room is equipped with ionization smoke detection and an automatic sprinkler system. The room has a floor area of over 17,000 square feet and a ceiling height of 23 feet nominally. Based on the installed defense-in-depth features, the size of the fire area, and the demonstrated performance time of 8 minutes, the NRC staff finds this OMA is acceptable for this specific room.

OMAs 1444, 1445, and 1448, for a fire in Rooms 772.0-A15 East, 772.0-A15 West or 772.0-A16, must be completed within 18 minutes. OMA 1448 applies to 772.0-A15 West only.

Demonstration of comparable actions resulted in a demonstrated time of less than 2 minutes. This provides approximately 16 minutes of margin for these actions. The fire area is equipped with a fire detection and automatic sprinkler system. The room has a floor area of 2153 square feet and a nominal ceiling height of 13 feet. Based on the installed defense-in-depth features, the size of the fire area, and the demonstrated performance time of 2 minutes, the NRC staff finds these OMAs are acceptable for these specific rooms.

OMAs 1495 and 1496, for a fire in Room 772.0-A15 West, must be completed within 20 minutes. Demonstration of comparable actions resulted in a demonstrated time of less than 4 minutes. This provides approximately 16 minutes of margin for these actions. The fire area is equipped with fire detection and automatic suppression systems. The room has a floor area of 2153 square feet and a nominal ceiling height of 13 feet. Based on the installed defense-in-depth features, the size of the fire area, and the demonstrated performance time of 4 minutes, the NRC staff finds these OMAs are acceptable for this specific room.

OMAs 1516 and 1517, for a fire in Room 757.0-A21, must be completed within 20 minutes. Demonstration of similar actions, which have no preceding actions, for Unit 1 indicated a travel and performance time of less than 3 minutes. This provides approximately 17 minutes of margin for these actions. The fire area is equipped with fire detection and an automatic sprinkler system. The area has a floor area of 2244 square feet with a nominal ceiling height of 14 feet. Based on the installed defense-in-depth features and the demonstrated performance of the OMA in approximately 3 minutes, the NRC staff finds these OMAs are acceptable for this specific room.

OMAs 1535 and 1536, for a fire in Room 737.0-A1N, must be completed within 20 minutes. Demonstration of similar actions, which have no preceding actions, for Unit 1 indicated a travel and performance time of 3 minutes. This provides 17 minutes of margin for these actions. The fire area is equipped with fire detection and an automatic sprinkler system. The area has a floor area of 23,144 square feet with a nominal ceiling height of 19 feet. Based on the installed defense-in-depth features and the demonstrated performance of the OMAs in approximately 3 minutes, the NRC staff finds these OMAs are acceptable for this specific room.

OMAs 1540 and 1542, for a fire in Room 737.0-A1B, must be completed within 20 minutes. Demonstration of similar actions, including preceding actions, for Unit 1 indicated a travel and performance time of approximately 3 minutes. This provides 17 minutes of margin for these actions. The fire area is equipped with fire detection and an automatic sprinkler system. The area has a floor area of 23,144 square feet with a nominal ceiling height of 19 feet. Based on the installed defense-in-depth features and the demonstrated performance of the OMAs in approximately 3 minutes, the NRC staff finds these OMAs are acceptable for this specific room.

Conclusion – Unit 2 Manual Actions

The NRC staff reviewed the submitted information regarding these specific OMAs and the fire scenarios that would cause them to be performed. The NRC staff concludes that, based on the fire protection defense-in-depth features and the feasibility and reliability of the OMAs, performance of these manual actions provides reasonable assurance that the capability to safely shutdown will be available, and is, therefore, acceptable.

6.1.10 Evaluation – Fire Hazards Analysis in Lieu of 10 CFR 50, Appendix R, Section III.G.2 Separation

In FPR Part VII, Section 2.9, TVA stated that there are rooms at WBN that lack the separation required by Section III.G.2 of Appendix R to 10 CFR Part 50. For these rooms, TVA relied upon a fire hazards analysis and an analysis of the safe shutdown capability rather than OMAs. In many cases these rooms are part of larger fire areas.

For all the rooms included as part of this evaluation, transient combustibles and ignition sources have been reported by TVA to be controlled by plant procedures. TVA provided a justification why certain ignition sources were not considered credible ignition sources. In addition, separation between adjacent rooms has been evaluated and TVA concluded that no credible fire could spread either from or to adjacent rooms. TVA reported that room fires affecting the FSSD equipment would neither initiate nor require a plant trip.

6.1.10.1   Rooms without Credible Ignition Sources and Redundant Trains

- Rooms 692.0-A29 and 692.0-A30 – Boric Acid Evaporator Package Rooms A and B
- Rooms 729.0-A1 and 737.0-A6 – Unit 1 South Main Steam Valve Room and Air Lock
- Room 729.0-A2 – Unit 1 North Main Steam Valve Room
- Room 729.0-A6 – Nitrogen Storage Area
- Room 729.0-A10 – Unit 2 North Main Steam Valve Room
- Rooms 729.0-A11 and 737.0-A10 – Unit 2 South Main Steam Valve Room and Air Lock
- Room 729.0-A12 – Unit 1 Steam Valve Instrument Room A
- Room 729.0-A13 – Unit 2 Steam Valve Instrument Room A
- Rooms 729.0-A15 and 763.5-A2 Upper Head Injection Equipment Rooms

TVA evaluated fire protection defense-in-depth for these rooms. These rooms have been reported to have minimal combustible loading consisting of plastics associated with small components or grease and oil associated with valves. Cables related to FSSD cables are installed within these rooms within conduit. Air lines that have a related FSSD function may be installed within these areas and are of welded steel construction. Other than cables within conduit and welded steel air piping, no other FSSD equipment is installed in these rooms. TVA evaluated the installed equipment in these rooms and concluded that there are no credible in situ ignition sources. Other ignition sources and transient combustibles are controlled in accordance with plant procedures. TVA determined that, even without any installed fire detection or suppression, no fire scenarios could credibly affect the cables or air lines that are involved in plant safe shutdown. TVA has determined that for each of these areas, if fire damage were to occur to the installed equipment a plant trip would not be initiated or required.

The NRC staff reviewed the submitted deviation and concludes that, based on the fire protection defense-in-depth features, limited combustibles and ignition sources, combustible controls, and no fires affecting FSSD equipment that would either initiate or require a plant trip, the configurations for these specific features in these rooms is acceptable to meet the underlying purpose of the rule and is, therefore, acceptable.

6.1.10.2   Room 757.0-A13 – Refueling Floor and New Fuel Storage Vault

The refueling floor has two fixed ignition sources installed, specifically two auxiliary air compressor units and equipment associated with hydraulic cranes and hoists. The air compressors, although credible ignition sources, are more than 20 feet separated from each other with no intervening combustibles. Therefore, a fire affecting one compressor would not be expected to affect the other compressor. A failure of one of the compressors could cause the air supply system to lose supply pressure. The other train would be available. In addition, a low pressure alarm on the affected system would be annunciated in the MCR.

The crane and hoist are only in operation when plant personnel are operating them. Therefore, any fire would be quickly identified by personnel in the immediate vicinity, and this would provide assurance that other FSSD equipment would not be damaged.

The new fuel storage vault has negligible combustibles and no credible ignition sources.

The NRC staff reviewed the submitted deviation and concludes that, based on the fire protection defense-in-depth features, limited combustibles, combustible controls, separation between redundant trains within the room with no intervening combustibles, continuous staffing when cranes and hoists are used, and no fires affecting FSSD equipment that would either initiate or require a plant trip, the configurations for these specific features in this room is acceptable to meet the underlying purpose of the rule and is, therefore, acceptable.

6.1.10.3   Room 757.0-A14 – Unit 2 Reactor Building Access Room and Room 757.0-A15 – Unit 2 Reactor Building Equipment Hatch

In contrast to the other rooms evaluated, these rooms have more than minimal combustible loading. The combustible loading is composed primarily of thermoset cable. The electrical circuits in the cables have circuit protection that reduces the likelihood of a self-ignited cable fire. TVA reported that there are no credible ignition sources in these rooms. In addition, each of these rooms is equipped with fire detection and automatic fire suppression systems.

For each of these rooms TVA identified five sets of redundant components. Each set of components is discussed below.

SGs 2 and 3 Main Steam Isolation Valves – The main steam isolation valves (MSIVs) are normally energized and fire damage that deenergizes the train will cause the MSIVs to close. Closed is the normal safe shutdown configuration. The fire damage failure mode of concern is a sustained hot short that keeps the MSIVs open.

In the unlikely event that damage causes a sustained hot short, given the limited ignition sources, full area detection and automatic suppression, the main steam system can be isolated from the MCR using the steam load valves.

RCP Seal Injection – An instrument cable for control circuits for the valve that controls the charging flow is located in these rooms near the ceiling. Based on the limited ignition sources and installation of an automatic fire suppression system, fire damage at the ceiling of these rooms is unlikely. In the unlikely event that the control circuits are damaged and the control valve spuriously operates, the indication is available and MCR operators could operate the valve using a different pressurizer level input or manually.

Control Cable for SG 3 PORV – A control cable for SG 3 PORV is routed through these rooms. A hot short to the control cable would cause the PORV to close, and not to be used for safe shutdown. In the unlikely event that a fire were to start, given the limited ignition sources, and the fire was not extinguished by the installed fire suppression system, the location of the cable in conduit over 20 feet above the floor provides assurance that cable damage would not occur.

Main Feedwater Isolation for SGs 2 and 3 – Main feedwater isolation valve control cables are installed in conduit in these rooms. Fire damage to these cables could interfere with the isolation of main feed water. In the unlikely event that a fire were to start, given the limited ignition sources, and the fire was not extinguished by the installed fire suppression system, operators in the MCR would still have available indication and controls over other valves that would be available to isolate the main feedwater flow.

Main Feedwater Bypass Line Isolation Valve Circuits for SGs 2 and 3 – Main feedwater bypass lines could remain open upon concurrent hot shorts of the control cables. In the unlikely event that a fire were to start, given the limited ignition sources, and fire was not extinguished by the installed fire suppression system, the control valves could still be closed by operator actions from the MCR.

The NRC staff reviewed the submitted deviation and concludes that, based on the fire protection defense-in-depth features, limited ignition sources, available detection and suppression systems, and either cables located high above the floor or alternative ways of meeting the safe shutdown goals using MCR actions, the configurations for these specific features in these rooms is acceptable to meet the underlying purpose of the rule and is, therefore, acceptable.

6.1.10.4   Unit 2 Containment Rooms

- Room 2RIR – Unit 2 Reactor Instrument Room
- Rooms 2RA1, 2RA2, 2RA3, and 2RA4 – Unit 2 Accumulator Rooms 1, 2, 3, and 4
- Rooms 2RF1 and 2RF2 – Unit 2 Reactor Building Fan Rooms 1 and 2
- Rooms 2RI-1, 2RI-2, 2RI-3, and 2RI-4 – Unit 2 Reactor Building Inside Crane Wall Rooms
- Rooms 2RO-1, 2RO-2, 2RO-3, and 2RO-4 – Unit 2 Reactor Building Outside Crane Wall Rooms

TVA stated that these rooms have stronger combustible controls than other plant areas, since these areas are considered combustible control zones. In addition, many of these areas are inaccessible during power operations and involve the climbing of ladders for entry, which will reduce the likelihood of transient combustibles and ignition sources. TVA stated that none of these rooms have credible in situ ignition sources. TVA provided a discussion that concluded fires in adjoining rooms would not affect the FSSD equipment in these rooms, due to either lack of combustibles in adjoining rooms or installed automatic suppression and detection in the adjoining rooms. TVA stated that a fire in one of these rooms affecting FSSD equipment would neither initiate nor require a plant trip.

In addition to defense-in-depth features described above, the FSSD capability has one or more of the additional features that provide(s) additional assurance that a fire in one of these rooms will not challenge plant safe shutdown:

- Redundant cables are separated by at least 3 feet horizontally,
- Cables are installed in conduit,
- Alternative systems are available in the control room to shutdown the plant,
- Spurious actuations are avoided by the use of dedicated conduit with no other energized conductors,
- Spurious actuations are avoided since they would only occur if there were a proper polarity two or three phase hot short,
- Targets are high above the floor, at least 10 feet, and/or
- Redundant trains may be located in the analysis volume, but not in the room being evaluated.

Based on its review of the information submitted by TVA, the NRC staff concludes that the lack of separation in these rooms, is an acceptable deviation from Section III.G.2.d of Appendix R to 10 CFR Part 50, because of the limited combustibles and ignition sources, failure of the FSSD equipment or cables would not initiate or require a plant trip, and all redundant safe shutdown circuits have one or more of the additional criteria above.

6.2   Deviations and Evaluations Related to BTP (APSCB) 9.5-1, Appendix A Guidance

6.2.1   Deviation – Fire Detection in Refueling Room and New Fuel Storage Vault

TVA committed to the guidance in Positions F.12 and F.13 of Appendix A to BTP (APCSB) 9.5-1, which states that fire detectors should be installed in new fuel and spent fuel pool areas. Contrary to the guidance, the refueling room (Room 757.0-A13), which includes the New Fuel Storage Vault (elevation 741.5 feet), is not provided with a detection system.

TVA states that the refueling room is constructed of reinforced concrete. This room has a large open area with a floor area of approximately 16,000 square feet and a nominal ceiling height of 56 feet. The walls, floor and penetration seals have a fire resistance rating of 2 hours or greater. The doors are not UL listed doors, but have been evaluated as equivalent to fire rated doors as listed in the FPR Part II, Table 14.8.1 (Fire Doors). The dampers have a minimum rating of 2 hours.

During normal operations, the in situ combustible loading in the refueling room and the new fuel storage vault is insignificant, resulting in an equivalent fire severity of less than 5 minutes. There are no ignition sources in the new fuel storage vault. The combustible materials in the refueling room are widely dispersed, which further diminishes the magnitude of a postulated fire. The combustibles consist of InstaCote (a plastic type fuel transfer canal coating); lube oil in air compressors; hoists and cranes; plastics associated with the electrical equipment, panels, fuel pool boundary, lighting and boxes; rubber fire hose; and anticipated amounts of radwaste trash and laundry. TVA further stated that transient combustibles in the room are controlled by WBN procedure NPG-SPP-18.4.7, "Control of Transient Combustibles." The potential ignition sources in the room are panels, air compressors, transformers, and lighting cabinets. The only ignition sources that could impact a FSSD component or cable are the Train A and B auxiliary air compressors.

The room is manned during an outage, which can assist in the early detection of a fire. The new fuel storage vault is only accessible from the refueling room and that access is normally closed with a steel hatch cover. The cover is removed when new fuel is received and stored until needed for a refueling outage. Due to the high ceiling and limited amount of combustibles, a fire in this area may not have sufficient energy to create the necessary air currents to carry the smoke to the ceiling. In this situation, the smoke detectors at the ceiling level may not be able to provide early detection in the event of a fire.

Standpipe and hose stations are provided in the refueling room and in adjacent rooms.

The Train A and B auxiliary air compressors supply backup air to the Train A and B air header if the normal air supply from the station air compressors is unable to maintain minimum pressure on the air header. A fire involving either of the auxiliary air compressors would not impact the normal air supply or the other auxiliary air compressor. The worse case fire scenario would be a loss of one train of auxiliary control air, which would not require either unit to shutdown. The other FSSD circuits are routed in conduits in the refueling floor area and are outside the fire zone of influence of the compressors. Therefore, a fire in the refueling room will not impact FSSD capability.

Based on its review of the information submitted by TVA, the NRC staff concludes that the lack of fire detection in the refueling room, including the new fuel storage vault, as identified above, is an acceptable deviation to the guidance of Positions F.12 and F.13 of Appendix A to BTP (APCSB) 9.5-1, because of the size of the refueling room, the limited amounts of in situ and transient combustibles, the separation of the room from other plant areas by fire-rated barriers, and the routing of FSSD circuits in conduits away from credible ignition sources.

6.2.2   Deviation – Fire Doors

TVA committed to the guidance in Position D.1.j in Appendix A to BTP (APCSB) 9.5-1, which states that door openings should be protected with equivalently rated fire doors, frames, and hardware that have been tested and approved by a nationally recognized laboratory.

In FPR Part VII, Section 4.1, TVA stated that, contrary to the guidance, a number of fire doors have been altered by the addition of signs and security hardware, or have been damaged and repaired onsite. Additionally, special-purpose doors, such as flood doors and pressure doors, are not UL labeled.

The fire doors that are not listed or labeled as fire-rated assemblies have been evaluated to the guidance of NFPA 80-1975, "Fire Doors and Windows," by TVA or nationally recognized laboratories for fire door assemblies. The evaluation criteria for fire door assemblies is documented and controlled by WBN General Engineering Specification-73, "Installation, Modification and Maintenance of Fire Protection Systems and Features."

FPR Part II, Table 14.8.1 lists the plant fire doors and the doors' fire-rating in hours. The table identifies doors that are not UL listed as having been evaluated and identified as equivalent to fire rated doors or they have been evaluated as being acceptable. A number of the fire doors at WBN have been altered by the addition of signs and security hardware or have been damaged and repaired. Examples of other fire doors that are not UL rated are special purpose doors such as flood doors and pressure doors, security doors in the MCR that are constructed of heavy welded steel construction and hollow core metal swinging doors.

Based on its review of the information submitted by TVA, the NRC staff concludes that TVA has adequately justified that certain door assemblies which do not fully meet the guidance in Position D.1.j in Appendix A to BTP (APCSB) 9.5-1 are acceptable, because the doors were evaluated to the guidance of NFPA 80-1975, and WBN General Engineering Specification-73 controls and documents the installation, modifications and maintenance of fire doors.

6.2.3   Deviation – Openings in Fire Walls

TVA committed to the guidance in Section D.1.j of Appendix A to BTP (APCSB) 9.5-1, which states that fire barriers should be capable of withstanding the fire hazards to which they could be exposed. NRC generic letters and guidance documents state that penetrations in walls, floors, and ceilings forming part of a fire barrier should be protected with seals or closure devices having a fire resistive rating equivalent to that required of the barrier.

In FPR Part VII, Section 4.2, TVA stated that there is a 6-inch wide by 3-inch deep gutter that penetrates each stairwell enclosure (Stairwells C1 and C2) from the corridor (Room 692.0-C11) in the control building.

These two stairwells are located at the opposite ends of the corridor (approximately 70 feet apart). The gutter penetrates the walls separating the stairwells from the corridor. Located in the gutter, there is one floor drain in each stairwell and two floor drains in the corridor.

The in situ combustible loading for the corridor is low and results in an equivalent fire severity of less than 20 minutes. The corridor is provided with a pre-action sprinkler system that is actuated by an ionization detection system. Standpipe and hose stations are in the two stairwells and portable extinguishers are provided in the corridor.

The in situ combustible liquids on elevation 692.0 feet of the control building are 35 gallons of lube oil associated with each of the two electrical board room chiller packages. The chiller packages are located in the Unit 2 mechanical equipment room, which is not part of Stairwells C1 or C2 or the corridor. However, the room is separated from Stairwell C2 by a 2-hour reinforced concrete wall. The combustibles in the Unit 2 mechanical equipment room consist of lube oil in the chillers, plastics associated with the electrical panels, boxes, lights and insulation on piping. The in situ combustible loading in this room is low resulting in an equivalent fire severity of less than 5 minutes. This room also has full detection and suppression installed.

Based on its review of the information submitted by TVA, the NRC staff concludes that, because of the low in situ combustible loading in the corridor, installed fire detection and suppression systems, available standpipe and hose stations in the two stairwells and portable extinguishers in the corridor, this deviation from Position D.1.j of Appendix A to BTP (APCSB) 9.5-1 for the corridor gutter that penetrate Stairwells C1 and C2 on control building elevation 692 feet is acceptable.

6.2.4   Deviation – Manual Hose Stations

TVA committed to the guidance in Section E.3.d of Appendix A to BTP (APCSB) 9.5-1, which states that interior manual hose installations should be able to reach any location with at least one effective hose stream. To accomplish this, standpipes with hose connections equipped with a maximum of 75 feet of 1-1/2 inch woven jacket lined fire hose and suitable nozzles should be provided.

In FPR Part VII, Section 4.3, TVA stated that there are manual hose stations with more than 75 feet of 1-1/2 inch UL listed or FM-approved fire hose located throughout the plant. The pressure loss in fire hoses due to conditions such as friction with the inner wall of the hose and turbulent water flow is directly proportional to the length of the hose. If the pressure loss is excessive the hose stream may not be effective.

To justify the use of hoses of greater than 75 feet in length up to 100 feet in length, TVA stated that these installations are consistent with the guidelines of NFPA 14-1974, "Standard for the Installation of Standpipe and Hose Systems," which allow up to 100 feet of hose connected to the standpipe.

For hose stations with more than 100 feet of hose, TVA stated that although those specific hose stations may not have been tested, hose stations at a higher elevation in the respective buildings were tested at a minimum of 65 psig at 500 gpm at a 2.5-inch hose connection. Also, TVA has calculated that there is 6 psi additional pressure loss for each additional 25 foot section of hose. TVA stated, in their letter dated May 30, 2012 (ADAMS Accession No. ML12153A374), that the tested hose stations are 31.5 feet higher in elevation than the hose stations with the additional hose. TVA calculated that 31.5 feet of elevation equates to approximately 13.5 psig of additional pressure at the lower elevation. This additional pressure on lower elevations would provide sufficient additional pressure to compensate for the approximately 6 psi of pressure loss for each of the two additional hose sections, and therefore would provide sufficient pressure and flow to meet the requirements of NFPA 14-1974.

Based on its review of the information submitted by TVA, the NRC staff concludes that the hose stations that have more than 75 feet of hose, as identified above, are acceptable deviations to the guidance of Section E.3.d of Appendix A to BTP 9.5-1.

6.2.5   Deviation – Fire Barrier Penetration between Fuel Oil Transfer Pump Room and the Diesel Generator Building Corridor

TVA committed to the guidance in Position D.1.j of Appendix A to BTP (APCSB) 9.5-1, which states that penetrations in fire barriers, including conduits and piping, should be sealed or closed to provide a fire resistance rating at least equal to that of the fire barrier itself. The fire hazard in each area should be evaluated to determine barrier requirements.

In FPR Part VII, Section 4.6, TVA stated that the fire barrier separating the fuel oil transfer pump room (Room 742.0-D8) from the diesel generator building corridor (Room 742.0-D9) is a 2-hour rated fire barrier and has a penetration containing a steel box. This penetration is not a tested fire-rated penetration assembly.

The fire barrier separating the fuel oil transfer pump room and the corridor is constructed of 8-inch thick reinforced concrete block and is fire-rated for 2 hours. The annular gap between the block wall and the box is filled with concrete grout, but no sealant material is installed within the box. The box back (inside the fuel oil transfer pump room) is a steel plate. The front of the panel is a steel plate with cutouts for three metal junction boxes.

The in situ combustible loading of the fuel oil transfer pump room is approximately 3,730 Btu/ft^2 and is due to insulation on cables associated with panel 0-L-162, hand switches, an emergency lighting unit, and foam plastic insulation. The in situ combustible loading of the corridor is approximately 77,700 Btu/ft^2 of which approximately 96 percent is due to insulation on cables in cable trays. The other in situ combustibles are dispersed throughout the corridor and do not present a direct exposure hazard to the box. The corridor width at the panel is approximately

6 feet. The door into the 2B-B diesel generator is across from the box and the door to the fuel oil transfer pump room is next to the box. The end of the corridor is less than 6 feet from this door. TVA stated that this arrangement minimizes the probability of transient combustibles being stored near the box.

The fuel oil transfer pump room is provided with a fire detection system and a total flooding automatic CO_2 suppression system. The detection system alarms in the MCR and actuates the suppression system. The corridor is provided with a fire detection system and an automatic sprinkler system. The detection system alarms in the MCR and actuates the suppression system. Upon receipt of a detection alarm, the MCR staff notifies the site fire brigade for both rooms.

The top of the box is located approximately 13 feet below the ceiling. TVA stated that in light of this distance and the location of the box at the end of the corridor, the detection system should alarm the MCR and actuate the suppression system before a hot gas layer could challenge the box.

TVA stated that the fuel oil transfer pump room and the corridor (analysis volume AV-081B) do not contain components required for safe shutdown in the event of a fire in these rooms. The small amount of in situ combustibles and the lack of free floor space limit the quantity of transient combustibles, thereby limiting the severity of a postulated fire in the room. The failure of a fuel oil line or pump that resulted in a fire is addressed by the total flooding, automatic CO_2 suppression system that will also control a postulated transient fire until the fire brigade responds.

Based on its review of the information submitted by TVA, the NRC staff concludes that, based on the penetration configuration and installed fire detection and automatic suppression systems, this configuration is adequate to prevent the passage of flames, hot gases or water from the corridor to the fuel oil transfer pump room or vice versa, and therefore, this non-tested, non-fire-rated penetration assembly is an acceptable deviation to the guidance in Position D.1.j of Appendix A to BTP (APCSB) 9.5-1.

6.2.6   Deviation – Undampered Penetrations between the Unit 1 Pipe Gallery and the Unit 1 Annulus and the Unit 2 Pipe Gallery and the Unit 2 Annulus

In FPR Part VII, Section 3.2, TVA stated that the walls separating the Unit 1 pipe gallery (Room 713.0-A6) from the Unit 1 Annulus and the Unit 2 pipe gallery (Room 713.0-A19) from the Unit 2 Annulus are 3-hour rated fire barriers. The containment purge air system return and exhaust ducts penetrate these walls in three places. The penetrations are not provided with fire dampers.

TVA provided the following details regarding these configurations:

- The ducts are constructed of 0.25 inch thick steel plates and welded schedule 10 pipe.
- As described in TVA's letter dated October 28, 2011 (ADAMS Accession No. ML11306A090), in response to NRC question RAI FPR VII-32, the connection between the duct and the purge air system is protected by 3M M20A wrap.
- The ducts are rigidly attached to the concrete wall.

- The penetrations are not straight-through, instead the openings in the concrete wall are offset to provide radiation protection.
- The ducts have no openings in the pipe chase.
- There is automatic detection and suppression installed in the annuluses and pipe chases.
- The two annuluses and the areas under the ducts in the pipe chases are combustible control zones.

Based on its review of the information submitted by TVA, the NRC staff concludes that, based on the physical configuration, installed fire protection systems, and administrative controls, the lack of fire dampers in these penetrations is an acceptable deviation from the guidance in Position D.1.j of Appendix A to BTP (APCSB) 9.5-1.

6.2.7   Deviation – Openings in Fire Barriers

Section D.1.j of Appendix A to BTP (APCSB) 9.5-1, "Guidelines for Fire Protection for Nuclear Plants Docketed Prior to July 1, 1976," specifies that penetrations in walls, floors, and ceiling forming part of a fire barrier be protected with self-closure devices having a fire-resistive rating equivalent to that of the barrier.

6.2.7.1   Ventilation and Purge Air Room Ventilation Penetrations

In FPR Part VII, Section 2.6.1, TVA stated that the ventilation and purge air (VPA) rooms (Rooms 737.0-A5 and 737.0-A9), the post-accident sampling system (PAS) rooms (Rooms 729.0-A8 and 729.0-A9) and the nitrogen storage room (Room 729.0-A6) are separated by 2-hour fire rated barriers. The walls and floor of the VPA rooms are penetrated by HVAC ducts that pass from the PAS rooms, enter the VPA rooms and then exit into the PAS and nitrogen storage room. TVA stated that the ducts have no fire dampers, but they also have no openings into the VPA rooms. Additionally, one duct enters each VPA room from the nitrogen storage room and terminates at a normally closed isolation damper. The ducts are constructed from Schedule 40 carbon steel pipe. Pipe sleeves are provided where the ducts penetrate the barriers between the VPA rooms and the PAS rooms and nitrogen storage rooms. Further, the annular space between the sleeves and the ducts is sealed with a fire-rated seal.

TVA stated that each of these rooms contains safe shutdown equipment. TVA further stated that the VPA and PAS rooms have fire detection and automatic fire suppression systems installed, and the nitrogen storage room has ionization smoke detection. Standpipe and hose systems are available in adjacent rooms and portable extinguishers are also available for manual fire-fighting in these rooms.

TVA stated that the significant fire exposure to the ducts from the VPA rooms consists of charcoal filter units in each VPA room. TVA also stated that closed-head water-spray suppression systems are provided for the charcoal filters and are actuated by duct-mounted ionization smoke detectors.

TVA stated that the effect of a fire in the PAS rooms or the nitrogen storage room would be experienced in the VPA rooms in the form of radiant heat from hot gases passing through the ducts. In the VPA rooms, TVA stated that no fixed combustibles are located in the immediate vicinity of these ducts, and the ducts are separated from the nearest safe shutdown circuit by more than 20 feet.

Based on the limited fire hazard, the installed fire detection and automatic fire suppression systems, the special hazard protection for the charcoal filters, and the construction of the ducts, the NRC staff concludes that the ducts will remain in place until a fire is extinguished and that the absence of fire dampers will not lead to fire propagation from one room to another. Therefore, this duct configuration is an acceptable deviation from the guidance of Section D.1.j of Appendix A to BTP (APCSB) 9.5-1.

6.2.7.2   Scuppers

6.2.7.2.1   ERCW Pump Room

TVA committed to the guidance in Position D.1.j of Appendix A to BTP (APCSB) 9.5-1, which states that penetrations in fire barriers, including conduits and piping, be sealed or closed to provide a fire resistance rating at least equal to that of the fire barrier itself.

In FPR Part VII, Section 2.6.2.1, TVA stated that, contrary to Position D.1.j, on elevation 741.0 feet of the IPS, there are four scupper openings penetrating the fire wall between the ERCW pump rooms and traveling screen rooms.

The wall separating the redundant ERCW pumps and the wall separating the ERCW pumps from the traveling screen pumps are 3-hour fire-rated barriers with the exception of the four scupper openings. These scupper openings are located at the floor and provide drainage of rainwater from the ERCW pump rooms to the traveling screen wells. The floor slopes away from the ERCW pumps toward the scuppers so that a fire in one ERCW pump room will not propagate through the scuppers and jeopardize a redundant train of ERCW pumps.

The wall separating the ERCW pump rooms and traveling screen rooms is intended to protect the rooms from the radiant heat of an exposure fire. The roof is designed as a missile shield and has beams that will allow free air flow from a fire to dissipate heat to the outside environment.

ERCW Pump Rooms A and B have heat detectors installed over the ERCW pumps and standpipe and hose stations are accessible for manual fire-fighting activities. TVA stated that even though these rooms are not provided with suppression and full area detection, the fire area barrier ratings are sufficient given the combustible loadings in the area.

Based on its review of the information submitted by TVA, the NRC staff concludes that the scupper configuration for the wall separating the ERCW pump rooms from the adjacent traveling screen rooms is an acceptable deviation from the guidance in Position D.1.j of Appendix A to BTP (APCSB) 9.5.1.

6.2.7.2.2   Yard Duct Bank

TVA committed to the guidance in Position D.1.j of Appendix A to BTP (APCSB) 9.5-1 which states that penetrations in fire barriers, including conduits and piping, be sealed or closed to provide a fire resistance rating at least equal to that of the fire barrier itself.

In FPR Part VII, Section 2.6.2.1, TVA stated that contrary to Position D.1.j, there are scupper openings in the Train A and Train B yard duct banks that run from the auxiliary building to the IPS where they share a common wall in three manholes.

Manholes 1A and 1B, 2A and 2B, and 3A and 3B are used to access the Train A and Train B duct banks that connect the auxiliary building to the IPS. The Train A and Train B duct banks are separated by a 12-inch thick reinforced concrete wall at each pair of manholes. One manhole in each pair contains a sump pump and is connected to the other manhole by a 2-inch diameter scupper opening. There are no other openings in the common wall separating the Train A and Train B manholes.

Cable insulation is the only combustible material in the yard duct banks where they share a common wall. The sump pumps are the only equipment in the yard duct banks where they share a common wall.

TVA stated that a postulated fire in the cable insulation of one duct bank or in the sump pump will not propagate through the scupper openings to the adjacent duct bank due to the lack of continuity of combustible materials between duct banks.

Based on its review of the information submitted by TVA, the NRC staff concludes that the scupper openings in the yard duct banks is an acceptable deviation from the guidance in Position D.1.j of Appendix A to BTP (APCSB) 9.5.1.

6.2.7.3   Auxiliary Building Penetrations

In FPR Part VII, Section 2.6.3, TVA described the following unprotected openings in the auxiliary building:

- *Open Stairs and Hatches.* TVA stated that water curtains designed in accordance with NFPA 13-1974, Section 4-4.8.2, have been installed to protect the openings listed in FPR Part VII, Section 2.6.3.1.

- *Sheet Metal Ducts That Are Not Provided with Fire Dampers.* TVA stated that these ducts are constructed of minimum 22 gauge sheet metal, are securely fastened to the fire barrier with angle steel, and that automatic suppression and detection is provided on at least one side of the opening. Finally, TVA stated that the safe shutdown analysis considered these openings as unprotected and ensured that a fire on either side of the opening would not impact both paths of redundant safe shutdown components, cables, or equipment.

- *Round HVAC Ducts Constructed of Spiral Welded Pipe or Schedule 10 Piping.* TVA stated that these ducts are treated as normal mechanical penetrations with an appropriate fire rated mechanical penetration seals.

- *Spare Conduit Sleeves.* As described in FPR Part VII, Section 2.6.3.3, TVA stated that spare conduit sleeves which penetrate fire barriers are provided with approved sealant material, capped on each end with metal caps or plugs, or a combination of the two.

- *Unrated Steel Hatches into Monolithic Concrete Enclosures.* As described in FPR Part VII, Section 2.6.3.4, TVA stated that the monolithic enclosures in which the steel hatches are located are not open to other rooms on other elevations. Further, TVA stated that there are no safe shutdown cables or components within the monolithic enclosures.

Based its review of the information submitted by TVA, the NRC staff concludes that these configurations are acceptable deviations from the guidance of Section D.1.j of Appendix A to BTP (APCSB) 9.5-1.

6.2.7.4 Control Building Equipment Hatches to the Turbine Building

In FPR Part VII, Section 2.6.4, TVA stated that the mechanical equipment rooms in the control building (Rooms 692.0-C1 and 692.0-C10) are provided with equipment hatches in the ceiling separating them from the turbine building. The equipment hatches have flush fitting steel covers which are not fire rated. TVA stated that the covers are vital area boundaries with access control and security features attached to the undersides, to prevent inadvertent removal.

TVA stated that the covers do not form a water tight seal, but will limit any flammable and combustible liquid spills through the hatch openings into the control building mechanical equipment rooms. Seepage could occur around the perimeter where the covers are mounted to the floor and through the small diameter holes in the covers that are provided to facilitate their removal.

TVA stated that there are no safe shutdown components in the turbine building within 20 feet of the equipment hatches, so that a fire that spreads up into the turbine building will not impact FSSD capability. Further, TVA stated that the mechanical equipment rooms are provided with automatic detection and pre-action sprinkler systems, including sidewall heads in the vicinity of the hatches. TVA stated that the installed detection and suppression systems would control or extinguish postulated fires passing through the hatch covers prior to arrival of the fire brigade.

Based on its review of the information submitted by TVA on the cover configuration, separation between FSSD equipment, and installed fire protection systems, the NRC staff concludes that the hatch covers are acceptable deviations from the guidance of Section D.1.j of Appendix A to BTP (APCSB) 9.5-1.

6.2.8 Evaluation – Large Fire Dampers

TVA committed to the guidance in Position D.1.j of Appendix A to BTP (APCSB) 9.5-1, which states that fire dampers should be tested and approved by a nationally recognized laboratory and the tests shall bound the installed configurations. In FPR Part VII, Section 3.4, TVA stated that in a December 12, 1984, report, UL stated that the maximum sizes of dampers covered by their classification and follow-up service program are 90 inches wide by 72 inches high in multiple assemblies (maximum sections being 30 inches wide by 36 inches high) and that dampers exceeding this are not eligible to be labeled. Contrary to this, fire dampers 1-ISD-31-3807 and 2-ISD-31-3882 consist of four 24-inch wide and 24½-inch high damper sections resulting in an opening 98⅝ inches wide by 24½ inches high. This exceeds the UL rated damper width by 8⅝ inches.

TVA further stated that fire tests reports dated June 15 and July 19, 1984, document the results of tests conducted by UL for Ruskin (the damper manufacturer) on large size damper installations. The large damper configurations in the two tests (100 inches by 91 inches and 100 inches by 72 inches) both passed the 3-hour fire endurance acceptance criteria by remaining in place and not having an opening in the damper configuration. Both configurations, however, failed the hose stream test at the end of the 3-hour fire exposure. The report dated December 12, 1984, documented UL's evaluation of WBN's installation of the large dampers.

The large fire damper installations at WBN are constructed from individual damper sections which are smaller than the maximum allowed by UL. The UL listed assembly is three sections wide by two sections high, but the WBN configuration is one section high and four sections wide, thus making the assembly more rigid and less susceptible to buckling and twisting under actual fire conditions. Also, the individual damper sections are 24 inches wide by 24½ inches high, which are less than the UL allowable 30 inches wide by 36 inches high. The overall damper height is 24½ inches high, and the UL allowable height is 72 inches, when two 36 inch dampers are stacked.

In the December 12, 1984, report, UL indicated that the WBN dampers (98⅝ inches wide by 24½ inches high) should have significantly less buckling and twisting of the vertical mullions than the tested damper (91 inches wide and 72 inches high) noted in the June 15, 1984, report. UL also concluded that the large damper installations at WBN provide adequate protection for their HVAC penetration.

Based on its review of the information submitted by TVA, the NRC staff concludes that, based on the UL review, TVA has adequately justified the above non-tested fire dampers that do not fully meet the guidance in Position D.1.j of Appendix A to BTP (APCSB) 9.5-1; therefore, the deviation from the guidance is acceptable.

6.2.9 Evaluation – Emergency Diesel Generators 7 Day Storage Tanks

TVA committed to the guidance in Position F.10 of Appendix A to BTP (APCSB) 9.5-1, which states that diesel fuel oil tanks with a capacity of over 1100 gallons should not be located inside buildings containing safety-related equipment. If located inside such buildings, the tanks should be separated by 3-hour fire barriers. Buried tanks are considered to meet the 3-hour fire resistance requirements.

In FPR Part VII, Section 4.4, TVA stated that there are four 7-day (70,248 gallon) storage tank assemblies, one per diesel generator, that are almost entirely buried below the floor of the diesel building. The fuel oil storage assembly for each diesel generator consists of four interconnected tanks, each with its own man-way access openings, one at either end of the tank. There are a total of 16 man-way access openings to the tanks from the corridor, and four in each diesel generator room. The man-way access openings are the only portion of the tanks that are not buried underneath the floor of the diesel generator building.

Each man-way access opening is in a pit covered by a removable plate cover sitting over the top of the pit flush with the floor. The cover is 1/4 inch thick steel plate, secured to the top of the tank by eighteen (18) 1/2 inch bolts. There are three normally closed openings in the cover plate. Two of the openings are provided for fuel oil circulation, and the other is for taking fuel oil samples.

The Pipe Gallery and Corridor (Room 742.0-D9) and Diesel Generator Units 1A-A, 2A-A, 1B-B and 2B-B (Rooms 742.0-D4, D5, D6 and D7) are provided with full area detection and automatic suppression systems. The diesel generator units each have heat detectors and a total flooding CO_2 suppression system. Standpipe and hose stations are provided within the diesel generator building on both elevations, and there are also fire hydrants available in the yard. The Pipe Gallery and Corridor has smoke detectors and an automatic pre-action sprinkler system. A standpipe and hose station is provided in the Pipe Gallery and Corridor.

Fire affects on the emergency diesel generators and associated cables in the diesel generator building will not have an adverse affect on safe shutdown. The diesel generators are not credited for any fire in the diesel generator building. The diesel generator building is located remotely from other buildings containing equipment or cables needed for safe shutdown. This is because offsite power capabilities have been evaluated and determined not to be affected or required for a fire in the diesel generator building, including the corridor.

Based on its review of the information submitted by TVA, the NRC staff concludes that, based on the physical construction of the man-way access openings, the man-way access openings being the only portion of the tanks that are not buried, the installed detection and suppression systems installed, the diesel generators not being required for any fire in the diesel generator building, location of the diesel generator building, and offsite power capabilities not being affected by a diesel generator building fire, the man-ways not being totally buried is an acceptable deviation from the guidance in Position F.10 of Appendix A to BTP (APCSB) 9.5-1.

6.2.10 Evaluation – Fire Dampers in the VCT Room Doors

In FPR Part VII, Section 3.5, TVA stated that a fire damper in the door connecting each of the two VCT rooms with the associated pipe gallery has been changed from a blade-type to a curtain-type configuration. The new dampers are damper/sleeve assemblies, installed with the damper inside the doors. The sleeve extends a short distance on each side of the opening. The door was tested with the original damper, but not with the new damper.

TVA provided the following details regarding these configurations:

- The combustible loading in the immediate vicinity of the doors is insignificant.
- The new dampers are listed dampers.
- The rooms on both sides of the doors are provided with automatic fire detection and suppression.

Based on its review of the information submitted by TVA, the NRC staff concludes that, based on the physical configuration, installed fire protection systems, and limited combustibles in the area, the change in these fire dampers is an acceptable deviation from the guidance in Position D.1.j of Appendix A to BTP (APCSB) 9.5-1.

6.2.11 Evaluation – Plexiglass Windows in the Security Control Point Building on the Refueling Floor

TVA committed to the guidance in Position D.1.d in Appendix A to BTP (APCSB) 9.5-1, which states, in part, that interior finishes should be noncombustible or have a flame spread rating of 25 or less.

In FPR Part VIII, TVA stated that, contrary to the guidance, the windows in a security control point building (on the 757.0 feet elevation on the Refueling Floor) was built with plexiglass windows, which do not meet the flame spread criteria. TVA stated the following concerning the plexiglass windows:

- Based on operating experience at Sequoyah Nuclear Plant, (i.e., a near-miss incident), glass windows pose a safety concern.
- Available alternatives either do not meet the flame spread criteria, or are not sufficiently transparent.
- The plexiglass windows add an insignificant amount of combustibles to a large room.
- The plexiglass windows have no effect on the safe shutdown analysis.
- The building is not used for safe shutdown.

Based on its review of the information submitted by TVA, the NRC staff concludes that, because of the minimal amount of combustibles involved and the lack of an effect on safe shutdown, the presence of the plexiglass windows in the security control point building on the Refueling Floor is an acceptable deviation from the guidance in Position D.1.d of Appendix A to BTP (APCSB) 9.5-1.

6.3   Additional Engineering Evaluations

6.3.1   Relaxation of FPR Surveillance Frequencies for the Reactor Buildings' Equipment Hatches

FPR Part VII, Section 6.1, summarizes TVA's evaluation of relaxing the surveillance frequencies for fire protection features (smoke detectors, sprinklers, Thermo-Lag, penetration seals) from their regular schedules for the equipment hatches (Rooms 757.0-A11 and -A15). TVA stated that these actions will be performed during outages, because these areas are inaccessible high radiation areas while the associated unit is operating.

These rooms connect the refueling floor and the reactor buildings, and provide equipment access. TVA stated that the rooms are constructed of reinforced concrete and are provided with smoke detectors and automatic pre-action sprinkler systems. FPR Part VI states that the rooms' barriers are 3-hour fire-rated, with the exception of the blast door into the reactor building. TVA stated that these doors are of heavy metal construction that would prevent a fire from propagating from either the reactor building into the room or from the room into the reactor building. TVA further stated that combustible loading in the rooms is comprised of cable insulation, light covers, and Thermo-Lag (Room 757.0-A11 only), and that there are no ignition sources in the rooms during power operation.

Based on its review of the information submitted by TVA, the NRC staff concludes that the described change in surveillance frequencies is reasonable to meet the ALARA radiation exposure requirements in 10 CFR Part 20, "Standards for Protection against Radiation," and, therefore, is acceptable.

6.3.2   Relaxation of FPR Surveillance Requirements for Fire Dampers in High Radiation and
        Contaminated Areas

In FPR Part VII, Section 6.2, TVA evaluated the need to perform surveillance for fire dampers in
high radiation or contaminated areas. TVA evaluated the consequences of the failure of the
following fire dampers to close during a fire event: 0-ISD-31-3846, 0-ISD-31-3847, and
0-ISD-31-3848. TVA stated that these fire dampers are located in contaminated areas and are
considered to be inaccessible.

6.3.2.1   Fire Damper 0-ISD-31-3846

TVA stated that fire damper 0-ISD-31-3846 is located in a 24-inch diameter embedded duct that
starts at an embedded collector box located in the Fuel Transfer Canal wall and runs for 40 feet
where it exits the concrete wall of the ventilation and purge air room (Room 737.0-A5) and then
enters a large (64 inch by 54 inch) duct.

TVA also stated that there is no combustible hazard in the fuel transfer canal, and negligible
quantities of combustibles in the vicinity of the duct in the ventilation and purge air room. TVA
further stated that the room is provided with smoke detection and automatic suppression.
Finally, TVA stated that should a fire breach the walls of the duct in the ventilation and purge air
room, the fire would have to travel a distance of 40 feet to reach the fuel transfer canal.

Based on its review of the information submitted by TVA, the NRC staff concludes that,
because of the limited combustibles in each room, the distance the fire would have to reach the
other room, the automatic suppression installed in the ventilation and purge air room, and the
ALARA concern identified by TVA, not performing surveillance of this fire damper is consistent
with Interpretation 4, "Fire Area Boundaries," of GL 86-10, and therefore, is acceptable.

6.3.2.2   Fire Dampers 0-ISD-31-3847 and 0-ISD-31-3848

TVA stated that one of the fire dampers is located in a 24-inch diameter embedded duct that
starts at an embedded collector box located in the spent fuel pit wall, runs for approximately
5 feet where it exits the concrete wall, traverses a corridor, and penetrates the concrete wall of
the ventilation and purge air room, and then enters a large (58 inch by 54 inch) duct. TVA
stated that the other fire damper is located in a 30-inch diameter embedded duct that starts at
an embedded collector box located in the opposite wall of the spent fuel pit, runs for
approximately 80 feet where it exits the spent fuel pit wall (near the 24-inch duct), traverses the
corridor, and penetrates the wall of the ventilation and purge air room and enters the large duct.

TVA stated that both ducts are coated with 2-inches of fire protective material (Pyrocrete)
where they traverse the corridor. Further, TVA stated that there are no combustible hazards in
the spent fuel pit, and negligible quantities of combustibles in the vicinity of the ducts in the
corridor and near the ducts in the ventilation and purge air room. In addition, the corridor and
the ventilation and purge air room are provided with smoke detection and the ventilation and
purge air room is also provided with automatic suppression. Finally, TVA stated that should a
fire breach the walls of the ducts in the ventilation and purge air room, the fire would have to
travel a distance of 10 feet or 80 feet to reach the spent fuel pit, which is filled with water.

Based on its review of the information submitted by TVA, the NRC staff concludes that,
because of the limited combustibles in each room, the distance the fire would have to reach the

other room, the automatic suppression installed in the ventilation and purge air room, and the ALARA concern identified by TVA, not performing surveillance of these two fire dampers is consistent with Interpretation 4, "Fire Area Boundaries," of GL 86-10, and therefore, is acceptable.

6.3.3   Gap between Door and Frame for Fire Door W9

In FPR Part VII, Section 6.3, TVA stated that a portion of the gap between the door and frame of fire door W9 exceeds the maximum 3/16-inch clearance. TVA further stated that the fire door is located in the wall that separates the RCW pump deck from the Train A ERCW pump room. TVA stated the following concerning the environment of door W9:

- The RCW pump deck is open to the atmosphere on three sides and does not have a roof.
- The ERCW pump room does not have a roof.
- The nearest RCW pump is located 17 feet horizontally from the door and the bottom of the door is 13.5 feet above the RCW pump deck.
- The in situ combustible load of the RCW pump deck consists primarily of lube oil associated with the RCW pumps.
- There are no in situ combustibles located directly under the door and the stairs and landings prevent any appreciable quantities of transient combustibles from being stored under the door.
- The door opens into a labyrinth that does not contain any in situ combustibles, nor are transient combustibles stored in the labyrinth.

Based on its review of the information submitted by TVA, the NRC staff concludes that, because of the physical configuration that would prevent the formation of a hot gas layer, the distance to the nearest source of combustibles, and limited amount of combustibles in this area, exceeding the allowable door to frame distance for fire door W9, as described in the FPR, is consistent with Interpretation 4, "Fire Area Boundaries," of GL 86-10, and therefore, is acceptable.

6.3.4   Relaxation of FPR Surveillance Requirements for Penetration Seals in High Radiation and Contaminated Areas

In FPR Part VII, Section 6.4, TVA evaluated the need to perform surveillance for penetration seals in high radiation areas by evaluating the consequences of the failure of the penetration seals for each of the rooms. TVA stated that their evaluations considered the locations not inspected, the proximity of combustibles, and the construction features of the rooms on either side of the seals.

6.3.4.1   Spent Resin Tank Room (Room 692.0-A15)

TVA stated that the penetration seals of interest in Room 692.0-A15 are installed in the wall separating it from the pipe gallery and chase room (Room 692.0-A24), which is a 2-hour rated fire barrier of reinforced concrete construction. TVA stated that the penetration seals are accessible for surveillance inspection from Room 692.0-A24, however, they are not accessible for inspection from the spent resin tank room due to the radiation posting of the room.

TVA stated that there is no safe shutdown equipment in the spent resin tank room. FPR Part VI stated that the combustible loading in both rooms is insignificant. TVA also stated that there is smoke detection installed in Room 692.0-A24.

Based on its review of the information submitted by TVA, the NRC staff concludes that, because of the minimal amount of combustibles in each room, the lack of safe shutdown equipment or cables in the spent resin tank room, the automatic smoke detection installed in the pipe gallery and chase room, and the ALARA concern identified by TVA, performing surveillance of these penetration seals from only one side is consistent with Interpretation 4, "Fire Area Boundaries," of NRC GL 86-10, and therefore, is acceptable.

6.3.4.2   Waste Hold Up Tank Room (Room 674.0-A1)

TVA stated that the penetration seals of interest in Room 674.0-A1 are installed in the wall separating it from the RHR Pump Room 1A-A (Room 676.0-A11) which is a 2-hour fire rated barrier of reinforced concrete construction. TVA stated that the penetration seals are accessible for surveillance inspection from Room 676.0-A11; however, they are not accessible for inspection from Room 674.0-A1 due to the radiation posting of the room.

TVA stated that there is no safe shutdown equipment required for a fire in the auxiliary building installed in the waste hold up tank room. FPR Part VI stated that the combustible loading in both rooms is insignificant. TVA also stated that there is smoke detection installed in Room 676.0-A11.

Based on its review of the information submitted by TVA, the NRC staff concludes that, because of the limited amount of combustibles in each room, the lack of safe shutdown equipment or cables in the waste hold up tank room, the automatic smoke detection installed in the RHR Pump Room 1A-A, and the ALARA concern identified by TVA, performing surveillance of these penetration seals from only one side is consistent with Interpretation 4, "Fire Area Boundaries," of GL 86-10, and therefore, is acceptable.

6.3.4.3   Hold Up Tank Rooms A and B (Rooms 676.0-A2 and 676.0-A3)

TVA stated that Rooms 676.0-A2 and 676.0-A3 are separated from adjacent non-high radiation area rooms by 2- and 3-hour fire rated barriers of reinforced concrete construction. TVA stated that the penetration seals are accessible for surveillance inspection from these adjacent rooms. The penetrations are not accessible from inside the hold up tank rooms for surveillance inspection due to the radiation posting of the rooms.

TVA stated that there is no safe shutdown equipment installed in the hold up tank rooms, nor any equipment that could initiate a plant trip. FPR Part VI stated that the combustible loading in both rooms is insignificant. Additionally, TVA stated that all the adjacent rooms which contain cables or equipment needed for FSSD have installed smoke detection.

Based on its review of the information submitted by TVA, the NRC staff concludes that, because of the limited amount of combustibles in these rooms, the lack of safe shutdown equipment or cables in the hold up tank room, the automatic smoke detection installed in the adjacent rooms which contain FSSD equipment or cables, and the ALARA concern identified by TVA, performing surveillance of these penetration seals from only one side is consistent with Interpretation 4, "Fire Area Boundaries," of GL 86-10, and therefore, is acceptable.

6.3.4.4   Gas Decay Tank Rooms (Rooms 692.0-A3 and 692.0-A5)

TVA stated that Rooms 692.0-A3 and 692.0-A5 are separated from adjacent non-high radiation area rooms by 2- and 3-hour fire rated barriers of reinforced concrete construction. TVA stated that the penetration seals are accessible for surveillance inspection from these adjacent rooms. The penetration seals are not accessible for inspection from the gas decay tank rooms due to the radiation posting of the rooms.

TVA stated that there is no safe shutdown equipment installed in the gas decay tank rooms, nor any equipment that could initiate a plant trip. FPR Part VI stated that the combustible loading in both rooms is insignificant. Additionally, TVA stated that all the adjacent rooms which contain cables or equipment needed for FSSD have installed automatic smoke detection.

Based on its review of the information submitted by TVA, the NRC staff concludes that, because of the limited amount of combustibles in these rooms, the lack of safe shutdown equipment or cables in the gas decay tank rooms, the automatic smoke detection installed in the adjacent rooms that contain FSSD equipment or cables, and the ALARA concern identified by TVA, performing surveillance of these penetration seals from only one side is consistent with Interpretation 4, "Fire Area Boundaries," of NRC GL 86-10, and therefore, is acceptable.

6.3.4.5   Barriers between High Radiation Area Rooms (Rooms 676.0-A2, 676.0-A3, 692.0-A3 and 692.0-A5)

TVA stated that the barriers between Rooms 676.0-A2 and 676.0-A3, Rooms 676.0-A2 and 692.0-A3, and Rooms 692.0-A3 and 692.0-A5 are not accessible because of the high levels of radiation present in these rooms.

TVA stated that there is no safe shutdown equipment installed in any of these rooms, nor any equipment that could initiate a plant trip. FPR Part VI stated that the combustible loading in all the rooms is insignificant. Additionally, TVA stated that all the adjacent rooms which contain cables or equipment needed for FSSD have installed automatic smoke detection.

Based on its review of the information submitted by TVA, the NRC staff concludes that, because of the limited amount of combustibles in these rooms, the lack of safe shutdown equipment or cables in Rooms 676.0-A2, 676.0-A3, 692.0-A3 and 692.0-A5, the automatic smoke detection installed in the adjacent rooms which contain FSSD equipment or cables, and the ALARA concern identified by TVA, not performing surveillance of these penetration seals is consistent with Interpretation 4, "Fire Area Boundaries," of GL 86-10, and therefore, is acceptable.

6.3.5   Diesel Generator Building Lube Oil Storage Room Fire Doors

The lube oil storage room (Room 742.0-D2) is a 3-hour fire-rated compartment. The 3-hour fire rated doors are in the open position and close only when the thermal link above the door melts or the CO_2 suppression system for the room discharges. To conform to the guidelines of NFPA 30 and 80, these doors should be self-closing. At each opening, TVA installed hollow metal side-hinged doors, which are normally closed. TVA stated that these doors are similar to rated fire doors and are expected to prevent smoke and hot gases from a fire from passing through the opening until the fusible links melt or the fire suppression system actuates.

Based on its review of the information submitted by TVA, the NRC staff concludes that the fire door configuration in the lube oil storage room complies with Position D.1.j of Appendix A to BTP (APCSB) 9.5-1 and, therefore, is acceptable.

7.0   CONCLUSION

On the basis of its review of TVA's as-designed FPR and TVA's supplemental information as referenced by this evaluation, the NRC staff concludes that the fire protection program for WBN, with the exception of Unit 1 specific OMAs, meets 10 CFR 50.48(a) and GDC 3 of Appendix A to 10 CFR Part 50, and is consistent with Sections III.G, III.J, III.L, and III.O of Appendix R to 10 CFR Part 50 and Appendix A to BTP (APCSB) 9.5-1, May 1976, with properly justified deviations and exceptions. Therefore, the NRC staff finds the as-designed FPR acceptable, contingent on the completion of the confirmatory items identified in Section 8.0 of this evaluation **(Open items 140, 141, 142, and 143, Appendix HH)**. NRC approval of the Unit 1 OMAs is documented in SSER 18, October 1995, of NUREG-0847, "Safety Evaluation Report Related to the Operation of Watts Bar Nuclear Plant Units 1 and 2."

8.0   CONFIRMATORY ITEMS

#	Item Description
(140)	TVA to confirm to the NRC staff the completion of the Unit 2 OMA feasibility walkdowns.
(141)	TVA to confirm to the NRC staff the completion of the Multiple spurious operation scenario resolution actions for scenarios which only affect Unit 2.
(142)	TVA to confirm to the NRC staff the completion of the electrical coordination modifications.
(143)	TVA to confirm the as-built FPR aligns with as-designed FPR. Gaps to be submitted to the NRC for approval.

APPENDIX HH

WATTS BAR UNIT 2 ACTION ITEMS TABLE

This table provides a status of required action items associated of all open items, confirmatory issues, and proposed license conditions that the NRC staff has identified. Unless otherwise noted, the item references are to sections of this SSER. Items that are still open are listed first, and items that have been closed are listed second. Some numbers were not used in the sequential list. There are **53 items** still open and **75 items** that have been closed as of April 17, 2013.

Open Items				
Item	**Type**	**Action Required**	**Lead**	**Status**
(1)	CI	Review evaluations and corrective actions associated with a power assisted cable pull. (NRC safety evaluation dated August 31, 2009, ADAMS Accession No. ML092151155)	NRR	Open
(2)	CI	Conduct appropriate inspection activities to verify cable lengths used in calculations and analysis match as-installed configuration. (NRC safety evaluation dated August 31, 2009, ADAMS Accession No. ML092151155)	RII	Open
(5)	CI	Verify timely submittal of pre-startup core map and perform technical review. (TVA letter dated September 7, 2007, ADAMS Accession No. ML072570676)	NRR	Open
(6)	CI	Verify implementation of TSTF-449. (TVA letter dated September 7, 2007, ADAMS Accession No. ML072570676)	NRR	Open
(7)	CI	Verify commitment completion and review electrical design calculations. (TVA letter dated October 9, 1990, ADAMS Accession No. ML073551056)	RII	Open
(8)	CI	TVA should provide a pre-startup map to the NRC staff indicating the rodded fuel assemblies and a projected end of cycle burnup of each rodded assembly for the initial fuel cycle 6-months prior to fuel load. (NRC safety evaluation dated May 3, 2010, ADAMS Accession No. ML101200035)	NRR	Open
(9)	CI	Confirm that education and experience of management and principal supervisory positions down through the shift supervisory level conform to Regulatory Guide 1.8. (SSER 22, Section 13.1.3)	RII	Open
(10)	CI	Confirm that TVA has an adequate number of licensed and non-licensed operators in the training pipeline to support the preoperational test program, fuel loading, and dual unit operation. (SSER 22, Section 13.1.3)	RII	Open

(12)		TVA's implementation of NGDC PP-20 and EDCR Appendix J is subject to future NRC audit and inspection. (SSER 22, Section 25.9)	NRR	Open
(13)		TVA is expected to submit an IST program and specific relief requests for WBN Unit 2 nine months before the projected date of OL issuance. (SSER 22, Section 3.9.6)	NRR	Open
(16)		Based on the uniqueness of EQ, the NRC staff must perform a detailed inspection and evaluation prior to fuel load to determine how the WBN Unit 2 EQ program complies with the requirements of 10 CFR 50.49. (SSER 22, Section 3.11.2)	RII/NRR	Open
(17)		The NRC staff should verify the accuracy of the WBN Unit 2 EQ list prior to fuel load. (SSER 22, Section 3.11.2.1)	RII/NRR	Open
(23)		Resolve whether or not TVA's reasoning for not upgrading the MSIV solenoid valves to Category I is a sound reason to the contrary, as specified in 10 CFR 50.49(l). (SSER 22, Section 3.11.2.2.1; SSER 24, Section 8.1)	NRR	Open
(25)		Prior to the issuance of an operating license, TVA is required to provide satisfactory documentation that it has obtained the maximum secondary liability insurance coverage pursuant to 10 CFR 140.11(a)(4), and not less than the amount required by 10 CFR 50.54(w) with respect to property insurance, and the NRC staff has reviewed and approved the documentation. (SSER 22, Section 22.3)	NRR	Open
(26)		For the scenario with an accident in one unit and concurrent shutdown of the second unit without offsite power, TVA stated that Unit 2 pre-operational testing will validate the diesel response to sequencing of loads on the Unit 2 emergency diesel generators (EDGs). The NRC staff will evaluate the status of this issue and will update the status of the EDG load response in a future SSER. (SSER 22, Section 8.1)	NRR	Open
(30)		TVA should confirm that all other safety-related equipment (in addition to the Class 1E motors) will have adequate starting and running voltage at the most limiting safety related components (such as motor operated valves, contactors, solenoid valves or relays) at the degraded voltage relay setpoint dropout setting. TVA should also confirm that the final Technical Specifications are properly derived from these analytical values for the degraded	RII/NRR	Open

		voltage settings. (SSER 22, Section 8.3.1.2)		
(32)		TVA should provide to the NRC staff the details of the administrative limits of EDG voltage and speed range, and the basis for its conclusion that the impact is negligible, and describe how it accounts for the administrative limits in the Technical Specification surveillance requirements for EDG voltage and frequency. (SSER 22, Section 8.3.1.14)	NRR	Open
(33)	CI	TVA stated in Attachment 9 of its letter dated July 31, 2010, that certain design change notices (DCNs) are required or anticipated for completion of WBN Unit 2, and that these DCNs were unverified assumptions used in its analysis of the 125 Vdc vital battery system. Verification of completion of these DCNs to the NRC staff is necessary prior to issuance of the operating license. (SSER 22, Section 8.3.2.3; SSER 24, Section 8.1)	RII/NRR	Open
(35)		TVA should provide information to the NRC staff that the CCS will produce feedwater purity in accordance with BTP MTEB 5-3 or, alternatively, provide justification for producing feedwater purity to another acceptable standard. (SSER 22, Section 10.4.6)	NRR	Open
(37)	CI	The NRC staff will review the combined WBN Unit 1 and 2 Appendix C prior to issuance of the Unit 2 OL to confirm (1) that the proposed Unit 2 changes were incorporated into Appendix C, and (2) that changes made to Appendix C for Unit 1 since Revision 92 and the changes made to the NP-REP since Revision 92 do not affect the bases of the staff's findings in this SER supplement. (SSER 22, Section 13.3.2)	NSIR	Open
(38)	CI	The NRC staff will confirm the availability and operability of the ERDS for Unit 2 prior to issuance of the Unit 2 OL. (SSER 22, Section 13.3.2.6)	RII/NSIR	Open
(40)	CI	The NRC staff will confirm the adequacy of the emergency facilities and equipment to support dual unit operations prior to issuance of the Unit 2 OL. (SSER 22, Section 13.3.2.8)	RII/NSIR	Open
(41)	CI	TVA committed to (1) update plant data displays as necessary to include Unit 2, and (2) to update dose assessment models to provide capabilities for assessing releases from both WBN units. The NRC staff will confirm the adequacy of these items prior to issuance of the Unit 2 OL. (SSER 22, Section 13.3.2.9)	RII/NSIR	Open

(43)	CI	Section V of Appendix E to 10 CFR Part 50 requires TVA to submit its detailed implementing procedures for its emergency plan no less than 180 days before the scheduled issuance of an operating license. Completion of this requirement will be confirmed by the NRC staff prior to the issuance of an operating license. (SSER 22, Section 13.3.2.18)	NSIR	Open
(49)	CI	The NRC staff was unable to determine how TVA linked the training qualification requirements of ANSI N45.2-1971 to TVA Procedure TI-119. Therefore, the implementation of training and qualification for inspectors will be the subject of future NRC staff inspections. (NRC letter dated July 2, 2010, ADAMS Accession No. ML101720050)	RII	Open
(51)	CI	The implementation of TVA Procedure TI-119 will be the subject of NRC follow-up inspection to determine if the construction refurbishment program requirements are being adequately implemented. (NRC letter dated July 2, 2010, ADAMS Accession No. ML101720050)	RII	Open
(59)		The staff's evaluation of the compatibility of the ESF system materials with containment sprays and core cooling water in the event of a LOCA is incomplete pending resolution of GSI-191 for WBN Unit 2. (SSER 23, Section 6.1.1.4)	NRR	Open
(61)		TVA should provide information to the NRC staff to demonstrate that PAD 4.0 can conservatively calculate the fuel temperature and other impacted variables, such as stored energy, given the lack of a fuel thermal conductivity degradation model. (SSER 23, Section 4.2.2)	NRR	Open
(63)	CI	TVA should confirm to the NRC staff that testing prior to Unit 2 fuel load has demonstrated that two-way communications is impossible with the Eagle 21 communications interface. (SSER 23, Section 7.2.1.1)	RII	Open
(64)	CI	TVA stated that, "Post modification testing will be performed to verify that the design change corrects the Eagle 21, Rack 2 RTD accuracy issue prior to WBN Unit 2 fuel load." This issue is open pending NRC staff review of the testing results. (SSER 23, Section 7.2.1.1)	RII	Open
(66)	CI	TVA should clarify FSAR Section 9.2.5 to add the capability of the UHS to bring the nonaccident unit to cold shutdown within 72 hours. (SSER 23, Section 9.2.5)	NRR	Open
(67)	CI	TVA should confirm, and the NRC staff should verify, that the component cooling booster pumps for Unit 2 are above PMF level. (SSER 23, Section 9.2.2)	RII	Open

(69)	CI	The WBN Unit 2 RCS vent system is acceptable, pending verification that the RCS vent system is installed. (SSER 23, Section 5.4.5)	RII	Open
(70)		TVA should provide the revised WBN Unit 2 PSI program ASME Class 1, 2, and 3 Supports "Summary Tables," to include numbers of components so that the NRC staff can verify that the numbers meet the reference ASME Code. (Section 3.2.3 of Appendix Z of SSER 23)	NRR	Open
(73)	CI	The NRC staff will inspect to confirm that TVA has completed the WBN Unit 2 EOPs prior to fuel load. (SSER 23, Section 7.5.3)	RII	Open
(74)	CI	The NRC staff will verify installation of the acoustic-monitoring system for the power-operated relief valve (PORV) position indication in WBN Unit 2 before fuel load. (SSER 23, Section 7.8.1)	RII	Open
(75)	CI	The NRC staff will verify that the test procedures and qualification testing for auxiliary feedwater initiation and control and flow indication are completed in WBN Unit 2 before fuel load. (SSER 23, Section 7.8.2)	RII	Open
(79)		TVA should perform a radiated susceptibility survey, after the installation of the hardware but prior to the RM-1000 being placed in service, to establish the need for exclusion distance for the HRCAR monitors while using handheld portable devices (e.g., walkie-talkie) in the control room, as documented in Attachment 23 to TVA's letter dated February 25, 2011, and item number 355 of TVA's letter dated April 15, 2011. (SSER 23, Section 7.5.2.3)	NRR	Open
(80)		TVA should provide clarification to the staff on how TVA Standard Specification SS-E18-14.1 meets the guidance of RG 1.180, and should address any deviations from the guidance of the RG. (SSER 23, Section 7.5.2.3)	NRR	Open
(83)	CI	TVA should confirm to the NRC staff the completion of the data storm test on the DCS. (SSER 23, Section 7.7.1.4)	NRR	Open
(90)	CI	The NRC staff should verify that the ERCW dual unit flow balance confirms that the ERCW pumps meet all specified performance requirements and have sufficient capability to supply all required ERCW normal and accident flows for dual unit operation and accident response, in order to verify that the ERCW pumps meet GDC 5 requirements for two-unit operation. (SSER 23, Section 9.2.1)	RII/NRR	Open
(91)		TVA should update the FSAR with information describing how WBN Unit 2 meets GDC 5, assuming the worst case single failure and a LOOP, as provided in TVA's letter dated April 13, 2011.	NRR	Open

		(SSER 23, Section 9.2.1)		
(93)		TVA should confirm to the staff that testing of the Eagle 21 system has sufficiently demonstrated that two-way communication to the ICS is precluded with the described configurations. (SSER 23, Section 7.9.3.2)	RII	Open
(115)	CI	TVA should update the FSAR to reflect the information regarding design changes to be implemented to lower radiation levels as provided in its letter the NRC dated June 3, 2010. (SSER 24, Section 12.5)	NRR	Open
(117)	CI	TVA should update the FSAR to reflect the calculational basis for access to vital areas as provided in its letter dated February 25, 2011. (SSER 24, Section 12.7.1)	NRR	Open
(131)		TVA should review the EOP action level setpoint to account for the difference between core exit temperature readings for Unit 1 and Unit 2 and confirm the EOP action level setpoint to the NRC staff. (SSER 24, Section 7.7.1.9.5)	NRR	Open
(133)		In order to confirm the stability analysis of the sand baskets used by TVA in the WBN Unit 2 licensing basis, TVA will perform either a hydrology analysis without crediting the use of the sand baskets at the Fort Loudoun dam for the seismic dam failure and flood combination, or TVA will perform a seismic test of the sand baskets, as stated in TVA's letter dated April 20, 2011. TVA will report the results of this analysis or test to the NRC by October 31, 2011. (SSER 24, Section 2.4.10)	NRR	Open
(134)		TVA should provide to the NRC staff supporting technical justification for the statements in Amendment 104 of FSAR Section 2.4.4.1, "Dam Failure Permutations," page 2.4-32 (in the section "Multiple Failures") that, "Fort Loudoun, Tellico, and Watts Bar have previously been judged not to fail for the OBE (0.09 g). Postulation of Tellico failure in this combination has not been evaluated but is bounded by the SSE failure of Norris, Cherokee, Douglas and Tellico." (SSER 24, Section 2.4.10)	NRR	Open
(139)	CI	The results of the cost-benefit analysis required by 10 CFR Part 50, Appendix I, subsection II.D, should be provided in the WBN Unit 2 FSAR. Upon receipt of the updated FSAR, the NRC staff will confirm that the update has been made by TVA. (SSER 25, Section 11.3)	NRR	Open

(140)	CI	TVA to confirm to the staff the completion of the Unit 2 OMA feasibility walkdowns. (SSER 26, Appendix FF, Section 8.0)	NRR	Open
(141)	CI	TVA to confirm to the staff the completion of the multiple spurious operation scenario resolution actions for scenarios which only affect Unit 2. (SSER 26, Appendix FF, Section 8.0)	NRR	Open
(142)	CI	TVA to confirm to the staff the completion of the electrical coordination modifications. (SSER 26, Appendix FF, Section 8.0)	NRR	Open
(143)	CI	TVA to confirm the as-built FPR aligns with as-designed FPR. Gaps to be submitted to the NRC for approval. (SSER 26, Appendix FF, Section 8.0)	NRR	Open

Closed Items				
(3)	CI	Confirm TVA submitted update to FSAR section 8.3.1.4.1. (NRC safety evaluation dated August 31, 2009, ADAMS Accession No. ML092151155) Closed in SSER 24, Section 8.1.	NRR	Closed
(4)	CI	Conduct appropriate inspection activities to verify that TVA's maximum SWBP criteria for signal level and coaxial cables do not exceed the cable manufacturer's maximum SWBP criteria. (NRC safety evaluation dated August 31, 2009, ADAMS Accession No. ML092151155) Closed in Inspection Report 0500391/2012602, dated March 27, 2012, ADAMS Accession No. ML12087A324.	RII	Closed
(11)	CI	The plant administrative procedures should clearly state that, when the Assistant Shift Engineer assumes his duties as Fire Brigade Leader, his control room duties are temporarily assumed by the Shift Supervisor (Shift Engineer), or by another SRO, if one is available. The plant administrative procedures should clearly describe this transfer of control room duties. (SSER 22, Section 13.1.3) Closed in SSER 25, Section 13.1.3.	NRR	Closed
(14)		TVA stated that the Unit 2 PTLR is included in the Unit 2 System Description for the Reactor Coolant System (WBN2-68-4001), which will be revised to reflect required revisions to the PTLR by September 17, 2010. (SSER 22, Section 5.3.1) Closed in SSER 25, Section 5.3.1.	NRR	Closed
(15)		TVA should confirm to the NRC staff the completion of Primary Stress Corrosion Cracking (PWSCC) mitigation activities on the Alloy 600 dissimilar metal	NRR	Closed

		butt welds (DMBWs) in the primary loop piping. (SSER 22, Section 3.6.3) Closed in SSER 24, Section 3.6.3.		
(18)		Based on the extensive layup period of equipment within WBN Unit 2, the NRC staff must review, prior to fuel load, the assumptions used by TVA to re-establish a baseline for the qualified life of equipment. The purpose of the staff's review is to ensure that TVA has addressed the effects of environmental conditions on equipment during the layup period. (SSER 22, Section 3.11.2.2) Closed in Inspection Report 0500391/2011604, dated June 29, 2011, ADAMS Accession No. ML111810890.	RII/NRR	Closed
(19)		The NRC staff should complete its review of TVA's EQ Program procedures for WBN Unit 2 prior to fuel load. (SSER 22, Section 3.11.2.2.1) Closed in Inspection Report 0500391/2011604, dated June 29, 2011, ADAMS Accession No. ML111810890.	RII/NRR	Closed
(20)	CI	Resolve whether or not routine maintenance activities should result in increasing the EQ of the 6.9 kV motors to Category I status in accordance with 10 CFR 50.49. (SSER 22, Section 3.11.2.2.1; SSER 24, Section 8.1) Closed in Inspection Report 0500391/2011605, dated August 5, 2011, ADAMS Accession No. ML112201418.	RII/NRR	Closed
(21)		The NRC staff should confirm that the Electrical Penetration Assemblies (EPAs) are installed in the tested configuration, and that the feedthrough module is manufactured by the same company and is consistent with the EQ test report for the EPA. (SSER 22, Section 3.11.2.2.1) Closed in Inspection Report 05000391/2011607, dated September 30, 2011, ADAMS Accession No. ML112730197.	RII/NRR	Closed
(22)		TVA must clarify its use of the term "equivalent" (e.g., identical, similar) regarding the replacement terminal blocks to the NRC staff. If the blocks are similar, then a similarity analysis should be completed and presented to the NRC for review. (SSER 22, Section 3.11.2.2.1) Closed in SSER 24, Section 8.1.	NRR	Closed
(24)		The NRC staff requires supporting documentation from TVA to justify its establishment of a mild environment threshold for total integrated dose of less than 1×10^3 rads for electronic components such as semiconductors or electronic components containing organic material. (SSER 22, Section 3.11.2.2.1) Closed in SSER 24, Section 8.1.	NRR	Closed

(27)		TVA should provide a summary of margin studies based on scenarios described in Section 8.1 for CSSTs A, B, C, and D. (SSER 22, Section 8.2.2) Closed in SSER 24, Section 8.1.	NRR	Closed
(28)		TVA should provide to the NRC staff a detailed discussion showing that the load tap changer is able to maintain the 6.9 kV bus voltage control band given the normal and post-contingency transmission operating voltage band, bounding voltage drop on the grid, and plant conditions. (SSER 22, Section 8.2.2) Closed in SSER 24, Section 8.1.	NRR	Closed
(29)		TVA should provide information about the operating characteristics of the offsite power supply at the Watts Bar Hydro Plant (for dual-unit operation), including the operating voltage range, postcontingency voltage drops (including bounding values and post-unit trip values), and operating frequency range. (SSER 22, Section 8.2.2) (corrected version of Open Item 29 from SSER 22 Appendix HH) Closed in SSER 24, Section 8.1.	NRR	Closed
(31)		TVA should clarify the loading sequence as explained in its letter dated December 6, 2010, to the staff. TVA should clarify whether the existing statements in FSAR regarding automatic sequencing logic are correct. If the FSAR description is correct, TVA should explain how the EDG and logic sequencing circuitry will respond to a LOCA followed by a LOOP scenario. (SSER 22, Section 8.3.1.11) (corrected version of Open Item 31 from SSER 22 Appendix HH) Closed in SSER 24, Section 8.1	NRR	Closed
(34)	CI	TVA stated that the method of compliance with Phase I guidelines would be substantially similar to the current Unit 1 program and that a new Section 3.12 will be added to the Unit 2 FSAR that will be materially equivalent to Section 3.12 of the current Unit 1 FSAR. (SSER 22, Section 9.1.4) Closed in SSER 24, Section 9.1.4.	NRR	Closed
(36)		TVA should provide information to the NRC staff to enable verification that the SGBS meets the requirements and guidance specified in the SER or provide justification that the SGBS meets other standards that demonstrate conformance to GDC 1 and GDC 14. (SSER 22, Section 10.4.8) Closed in SSER 24, Section 10.4.8.	NRR	Closed
(39)	CI	The NRC staff will confirm the adequacy of the communications capability to support dual unit operations prior to issuance of the Unit 2 OL. (SSER 22, Section 13.3.2.6) Closed in Inspection Report 0500391/2011609, dated December 16,	RII/NSIR	Closed

		2011, ADAMS Accession No. ML11350A229.		
(42)	CI	The NRC staff will confirm the adequacy of the accident assessment capabilities to support dual unit operations prior to issuance of the Unit 2 OL. (SSER 22, Section 13.3.2.9) Closed in Inspection Report 0500391/2011609, dated December 16, 2011, ADAMS Accession No. ML11350A229.	RII/NSIR	Closed
(44)		TVA should provide additional information to clarify how the initial and irradiated RT_{NDT} was determined. (SSER 22, Section 5.3.1) Closed in SSER 25, Section 5.3.1.	NRR	Closed
(45)	CI	TVA stated in its response to RAI 5.3.2-2, dated July 31, 2010, that the PTLR would be revised to incorporate the COMS arming temperature. (SSER 22, Section 5.3.2) Closed in SSER 25, Section 5.3.2.	NRR	Closed
(46)	CI	The LTOP lift settings were not included in the PTLR, but were provided in TVA's response to RAI 5.3.2-2 in its letter dated July 31, 2010. TVA stated in its RAI response that the PTLR would be revised to incorporate the LTOP lift settings into the PTLR. (SSER 22, Section 5.3.2) Closed in SSER 25, Section 5.3.2.	NRR	Closed
(47)	CI	The NRC staff noted that TVA's changes to Section 6.2.6 in FSAR Amendment 97, regarding the implementation of Option B of Appendix J, were incomplete, because several statements remained regarding performing water-sealed valve leakage tests "as specified in 10 CFR [Part] 50, Appendix J." With the adoption of Option B, the specified testing requirements are no longer applicable; Option A to Appendix J retains these requirements. The NRC discussed this discrepancy with TVA in a telephone conference on September 28, 2010. TVA stated that it would remove the inaccurate reference to Appendix J for specific water testing requirements in a future FSAR amendment. (SSER 22, Section 6.2.6) Closed in SSER 26, Section 6.2.6.	NRR	Closed
(48)	CI	The NRC staff should verify that its conclusions in the review of FSAR Section 15.4.1 do not affect the conclusions of the staff regarding the acceptability of Section 6.5.3. (SSER 22, Section 6.5.3) Closed in SSER 26, Section 6.5.3.	NRR	Closed
(50)	CI	TVA stated that about 5 percent of the anchor bolts for safety-related pipe supports do not have quality control documentation, because the pull tests have not yet been performed. Since the documentation is still under development, the NRC staff will conduct	RII	Closed

		inspections to follow-up on the adequate implementation of this construction refurbishment program requirement. (NRC letter dated July 2, 2010, ADAMS Accession No. ML101720050) Closed in Inspection Report 0500391/2013612, dated March 28, 2013, ADAMS Accession No. ML13088A066.		
(52) through (58)		Not used.		
(60)	CI	TVA should amend the FSAR description of the design and operation of the spent fuel pool cooling and cleanup system in FSAR Section 9.1.3 as proposed in its December 21, 2010, letter to the NRC. (SSER 23, Section 9.1.3) Closed in SSER 26, Section 9.1.3.	NRR	Closed
(62)	CI	Confirm TVA's change to FSAR Section 10.4.9 to reflect its intention to operate with each CST isolated from the other. (SSER 23, Section 10.4.9) Closed in SSER 24, Section 10.4.9.	NRR	Closed
(65)		TVA should provide justification to the staff regarding why different revisions of WCAP-13869 are referenced in WBN Unit 1 and Unit 2. (SSER 23, Section 7.2.1.1) Closed in SSER 26, Section 7.2.1.1.	NRR	Closed
(68)		Not used.		
(71)		By letter dated April 21, 2011 (ADAMS Accession No. ML111110513), TVA withdrew its commitment to replace the Unit 2 clevis insert bolts. TVA should provide further justification for the decision to not replace the bolts to the NRC staff. (SSER 23, Section 3.9.5) Closed in SSER 26, Section 3.9.5.	NRR	Closed
(72)		The NRC staff should complete its review and evaluation of the additional information provided by TVA regarding the ICC instrumentation. (SSER 23, Section 4.4.8) Closed in SSER 25, Section 7.5.2.2.	NRR	Closed
(76)	CI	The NRC staff will verify that the derivative time constant is set to zero in WBN Unit 2 before fuel load. (SSER 23, Section 7.8.3) Closed in Inspection Report 05000391/2011607, dated September 30, 2011, ADAMS Accession No. ML112730197.	RII	Closed
(77)		It is unclear to the NRC staff which software V&V documents are applicable to the HRCAR monitors. TVA should clarify which software V&V documents are applicable, in order for the staff to complete its evaluation. (SSER 23, Section 7.5.2.3) Closed in SSER 26, Section 7.5.2.3.4	NRR	Closed

(78)		TVA intends to issue a revised calculation reflecting that the TID in the control room is less than 1×10^3 rads, which will be evaluated by the NRC staff. (SSER 23, Section 7.5.2.3) Closed in SSER 25, Section 7.5.2.3.	NRR	Closed
(81)		The extent to which TVA's supplier, General Atomics (GA), complies with EPRI TR-106439 and the methods that GA used for its commercial dedication process should be provided by TVA to the NRC staff for review. (SSER 23, Section 7.5.2.3) Closed in SSER 26, Section 7.5.2.3.4.	NRR	Closed
(82)		The staff concluded that the information provided by TVA pertaining to the in-containment LPMS equipment qualification for vibration was incomplete. TVA should provide (item number 362 of ADAMS Accession No. ML111050009), documentation that demonstrates the LPMS in-containment equipment has been qualified to remain functional in its normal operating vibration environment, per RG 1.133, Revision 1. (SSER 23, Section 7.6.1) Closed in SSER 24, Section 7.6.1.4.5.	NRR	Closed
(84) through (89)		Not used.		
(92)		Not used.		
(94)		TVA should provide to the staff either information that demonstrates that the WBN Unit 2 Common Q PAMS meets the applicable requirements in IEEE Std. 603-1991, or justification for why the Common Q PAMS should not meet those requirements. (SSER 23, Section 7.5.2.2.3) Closed in SSER 26, Section 7.5.2.2)	NRR	Closed
(95)		TVA should update FSAR Table 7.1-1, "Watts Bar Nuclear Plant NRC Regulatory Guide Conformance," to reference IEEE Std. 603-1991 for the WBN Unit 2 Common Q PAMS. (SSER 23, Section 7.5.2.2.3) Closed in SSER 25, Section 7.5.2.2.	NRR	Closed
(96)		TVA should (1) update FSAR Table 7.1-1 to include RG 1.100, Revision 3, for the Common Q PAMS, or (2) demonstrate that the Common Q PAMS is in conformance with RG 1.100, Revision 1, or provide justification for not conforming. (SSER 23, Section 7.5.2.2.3) Closed in SSER 25, Section 7.5.2.2.	NRR	Closed
(97)		TVA should demonstrate that the WBN Unit 2 Common Q PAMS is in conformance with RG 1.153, Revision 1, or provide justification for not conforming. (SSER 23, Section 7.5.2.2.3) Closed	NRR	Closed

		in SSER 25, Section 7.5.2.2.		
(98)		TVA should demonstrate that the WBN Unit 2 Common Q PAMS is in conformance with RG 1.152, Revision 2, or provide justification for not conforming. (SSER 23, Section 7.5.2.2.3) Closed in SSER 26, Section 7.5.2.2.3.	NRR	Closed
(99)		TVA should update FSAR Table 7.1-1 to reference IEEE 7-4.3.2-2003 as being applicable to the WBN Unit 2 Common Q PAMS. (SSER 23, Section 7.5.2.2.3; SSER 25, Section 7.5.2.2) Closed in SSER 25, Section 7.5.2.2.	NRR	Closed
(100)		TVA should update FSAR Table 7.1-1 to reference RG 1.168, Revision 1; IEEE 1012-1998; and IEEE 1028-1997 as being applicable to the WBN Unit 2 Common Q PAMS. (SSER 23, Section 7.5.2.2.3) Closed in SSER 25, Section 7.5.2.2.	NRR	Closed
(101)		TVA should demonstrate that the WBN Unit 2 Common Q PAMS application software is in conformance with RG 1.168, Revision 1, or provide justification for not conforming. (SSER 23, Section 7.5.2.2.3) Closed in SSER 26, Section 7.5.2.2.	NRR	Closed
(102)		TVA should update FSAR Table 7.1-1 to reference RG 1.209 and IEEE Std. 323-2003 as being applicable to the WBN Unit 2 Common Q PAMS. (SSER 23, Section 7.5.2.2.3) Closed in SSER 25, Section 7.5.2.2.	NRR	Closed
(103)		TVA should demonstrate that the WBN Unit 2 Common Q PAMS conforms to RG 1.209 and IEEE Std. 323-2003, or provide justification for not conforming. (SSER 23, Section 7.5.2.2.3) Closed in SSER 25, Section 7.5.2.2.	NRR	Closed
(104)	CI	The NRC staff will review the WEC self assessment to verify that it the WBN Unit 2 PAMS is compliant to the V&V requirements in the SPM or that deviations from the requirements are adequately justified. (SSER 23, Section 7.5.2.2.3.4.2) Closed in SSER 25, Section 7.5.2.2.	NRR	Closed
(105)		TVA should produce an acceptable description of how the WBN Unit 2 Common Q PAMS SysRS and SRS implement the design basis requirements of IEEE Std. 603-1991 Clause 4. (SSER 23, Section 7.5.2.2.3.4.3.1) Closed in SSER 26, Section 7.5.2.2.	NRR	Closed
(106)		TVA should produce a final WBN Unit 2 Common Q PAMS SRS that is independently reviewed. (SSER 23, Section 7.5.2.2.3.4.3.1) Closed in SSER 25, Section 7.5.2.2.	NRR	Closed

(107)	CI	TVA should provide to the NRC staff documentation to confirm that the final WBN Unit 2 Common Q PAMS SDDs that are independently reviewed. (SSER 23, Section 7.5.2.2.3.4.3.2) Closed in SSER 25, Section 7.5.2.2.	NRR	Closed
(108)		TVA should demonstrate to the NRC staff that there are no synergistic effects between temperature and humidity for the Common Q PAMS equipment. (SSER 23, Section 7.5.2.2.3.5.2) Closed in SSER 26, Section 7.5.2.2.	NRR	Closed
(109)		TVA should demonstrate to the NRC staff acceptable data storm testing of the Common Q PAMS. (SSER 23, Section 7.5.2.2.3.7.1.8) Closed in SSER 25, Section 7.5.2.2.	NRR	Closed
(110)		TVA should provide information to the NRC staff describing how the WBN Unit 2 Common Q PAMS design supports periodic testing of the RVLIS function. (SSER 23, Section 7.5.2.2.3.9.2.6) Closed in SSER 26, Section 7.5.2.2.	NRR	Closed
(111)		TVA should confirm to the staff that there are no changes required to the technical specifications as a result of the modification installing the Common Q PAMS. If any changes to the technical specifications are required, TVA should provide the changes to the NRC staff for review. (SSER 23, Section 7.5.2.2.3.11) Closed in SSER 26, Section 7.5.2.2.	NRR	Closed
(112)	CI	TVA should provide an update to the FSAR reflecting the radiation protection design features descriptive information provided in its letter dated October 4, 2010. (SSER 24, Section 12.4) Closed in SSER 26, Section 12.4.	NRR	Closed
(113)	CI	TVA should provide an update to the FSAR reflecting the justification for the periodicity of the COT frequency for WBN non-safety related area radiation monitors. (SSER 24, Section 12.4) Closed in SSER 26, Section 12.4.	NRR	Closed
(114)	CI	TVA should update the FSAR to reflect that WBN meets the radiation monitoring requirements of 10 CFR 50.68. (SSER 24, Section 12.4) Closed in SSER 26, Section 12.4.	NRR	Closed
(116)	CI	TVA should update the FSAR to reflect the qualification standards of the RPM as provided in its letter to the NRC dated October 4, 2010. (SSER 24, Section 12.6) Closed in SSER 26, Section 12.6.	NRR	Closed

(118)		TVA should provide to the NRC staff a description of how the other vanadium detectors within the IITA would be operable following the failure of an SPND. (SSER 24, Section 7.7.1.9.2) Closed in SSER 26, Section 7.7.1.9.	NRR	Closed
(119)		TVA should submit WNA-CN-00157-WBT, Revision 0, to the NRC by letter. The NRC staff should confirm by review of WNA-CN-00157-WBT, Revision 0, that no credible source of faulting can negatively impact the CETs or PAMS train. (SSER 24, Section 7.7.1.9.5) Closed in SSER 25, Section 7.7.1.9.	NRR	Closed
(120)		TVA must confirm to the NRC staff that the maximum over-voltage or surge voltage that could affect the system is 264 VAC, assuming that the power supply cable to the SPS cabinet is not routed with other cables greater than 264 VAC. (SSER 24, Section 7.7.1.9.5; SSER 25, Section 7.7.1.9) Closed in SSER 26, Section 7.7.1.9.	NRR	Closed
(121)		TVA should submit the results to the NRC staff of a 600 VDC dielectric strength test performed on the IITA assembly. (SSER 24, Section 7.7.1.9.5) Closed in SSER 26, Section 7.7.1.9.	NRR	Closed
(122)		TVA should confirm to the NRC staff that different divisions of safety power are supplied to the IIS SPS cabinets, with the power cables routed in separate shielded conduits. (SSER 24, Section 7.7.1.9.5) Closed in SSER 25, Section 7.7.1.9.	NRR	Closed
(123)		TVA should provide an explanation to the NRC staff of how the system will assign a data quality value to notify the power distribution calculation software to disregard data from a failed SPND. (SSER 24, Section 7.7.1.9.5) Closed in SSER 26, Section 7.7.1.9.	NRR	Closed
(124)		While the BEACON datalink on the Application server can connect to either BEACON machine, only BEACON A is used for communication. TVA should clarify to the NRC staff whether automatic switchover to the other server is not permitted. (SSER 24, Section 7.7.1.9.5) Closed in SSER 25, Section 7.7.1.9.	NRR	Closed
(125)		TVA should provide clarification to the NRC staff of the type of connector used with the MI cable in Unit 2, and which EQ test is applicable. (SSER 24, Section 7.7.1.9.5) Closed in SSER 26, Section 7.7.1.9.	NRR	Closed
(126)		To enable the NRC staff to evaluate and review the IITA environmental qualification, TVA should provide the summary report of the environmental qualification for the IITA. (SSER 24,	NRR	Closed

		Section 7.7.1.9.5) Closed in SSER 26, Section 7.7.1.9.		
(127)		TVA should provide a summary to the NRC staff of the electro-magnetic interference/radio-frequency interference (EMI/RFI) testing for the MI cable electro-magnetic compatibility (EMC) qualification test results. (SSER 24, Section 7.7.1.9.5) Closed in SSER 26, Section 7.7.1.9.	NRR	Closed
(128)		TVA should submit the seismic qualification test report procedures and results for the SPS cabinets to the NRC staff for review. (SSER 24, Section 7.7.1.9.5) Closed in SSER 25, Section 7.7.1.9.	NRR	Closed
(129)		TVA should verify to the NRC staff resolution of the open item in WNA-CN-00157-WBT for the Quint power supply (to be installed in the SPS cabinet) to undergo EMC testing of 4 kV to validate the assumptions made in the Westinghouse analysis. (SSER 24, Section 7.7.1.9.5) Closed in SSER 26, Section 7.7.1.9.	NRR	Closed
(130)		TVA should provide a summary to the NRC staff of the EMC qualification test results of the SPS cabinets. (SSER 24, Section 7.7.1.9.5) Closed in SSER 25, Section 7.7.1.9.	NRR	Closed
(132)		TVA must provide the NRC staff with analyses of the boron dilution event that meet the criteria of SRP Section 15.4.6, including a description of the methods and procedures used by the operators to identify the dilution path(s) and terminate the dilution, in order for the staff to determine that the analyses comply with GDC 10. (SSER 24, Section 15.2.4.4) Closed in SSER 26, Section 15.2.4.4.	NRR	Closed
(135)		TVA has not provided the analysis required by 10 CFR Part 50, Appendix I, subsection II.D. TVA must demonstrate with a cost-benefit analysis that a sufficient reduction in the collective dose to the public within a 50-mile radius would not be achieved by reasonable changes to the design of the WBN gaseous effluent processing systems. (SSER 24, Section 11.3) Closed in SSER 25, Section 11.3.	NRR	Closed
(136)	CI	The JFD summary for the data from 1991 through 2010 provided by letter dated November 7, 2011, and a discussion of the long-term representativeness of these data should be provided in the WBN Unit 2 FSAR. Upon receipt of the updated FSAR, the NRC staff will confirm that these updates have been made by TVA. (SSER 25, Section 2.3.3) Closed in SSER 26, Section 2.3.3.	NRR	Closed

(137)	CI	The NRC staff will confirm, upon receipt, that TVA integrated the updated CR χ/Q values from its letter dated September 15, 2011, into a future amendment of the FSAR. (SSER 25, Section 2.3.4) Closed in SSER 26, Section 2.3.4.	NRR	Closed
(138)	CI	Upon receipt of the updated ODCM, the NRC staff will confirm that corresponding revisions related to the updated annual average χ/Q and D/Q values have been made to the ODCM. (SSER 25, Section 2.3.5) Closed in SSER 26, Section 2.3.5.	NRR	Closed

CI – Confirmatory Issue

NRC FORM 335
(12-2010)
NRCMD 3.7

U.S. NUCLEAR REGULATORY COMMISSION

BIBLIOGRAPHIC DATA SHEET

(See instructions on the reverse)

1. REPORT NUMBER
(Assigned by NRC, Add Vol., Supp., Rev., and Addendum Numbers, if any.)

NUREG-0847
Supplement No. 26

2. TITLE AND SUBTITLE

Safety Evaluation Report
Related to the Operation of
Watts Bar Nuclear Plant, Unit 2
Docket No. 50-391

3. DATE REPORT PUBLISHED

MONTH	YEAR
June	2013

4. FIN OR GRANT NUMBER

5. AUTHOR(S)

J. Poole, et al.

6. TYPE OF REPORT

Technical

7. PERIOD COVERED (Inclusive Dates)

8. PERFORMING ORGANIZATION - NAME AND ADDRESS (If NRC, provide Division, Office or Region, U. S. Nuclear Regulatory Commission, and mailing address; if contractor, provide name and mailing address.)

U.S. Nuclear Regulatory Commission
Office of Nuclear Reactor Regulation
Division of Operating Reactor Licensing
Washington, DC 20555-0001

9. SPONSORING ORGANIZATION - NAME AND ADDRESS (If NRC, type "Same as above", if contractor, provide NRC Division, Office or Region, U. S. Nuclear Regulatory Commission, and mailing address.)

Same as above.

10. SUPPLEMENTARY NOTES
Docket No. 50-391

11. ABSTRACT (200 words or less)

This report supplements the safety evaluation report (SER), NUREG-0847 (June 1982), Supplement No. 25 (November 2011, Agencywide Documents Access and Management System (ADAMS) Accession No. ML12011A024), with respect to the application filed by the Tennessee Valley Authority (TVA), as applicant and owner, for a license to operate Watts Bar Nuclear Plant (WBN) Unit 2 (Docket No 50-391).

In its SER and Supplemental SER (SSER) Nos. 1 through 20 issued by the U.S. Nuclear Regulatory Commission (NRC) staff, the NRC staff documented its safety evaluation and determination that WBN Unit 1 met all applicable regulations and regulatory guidance. Based on satisfactory findings from all applicable inspections, on February 7, 1996, the NRC issued a full-power operating license (OL) to WBN Unit 1, authorizing operation up to 100-percent power.

In SSERs subsequent to SSER 20, the staff addressed TVA's application for a license to operate WBN Unit 2, and provided information regarding the status of the items remaining to be resolved, which were open at the time that TVA deferred construction of WBN Unit 2, and were not evaluated and resolved as part of the licensing of WBN Unit 1. In this and future SSERs, the NRC staff will document its evaluation and closure of open items in its review of TVA's application for an operating license for WBN Unit 2.

12. KEY WORDS/DESCRIPTORS (List words or phrases that will assist researchers in locating the report.)

Safety Evaluation Report (SER)
Watts Bar Nuclear Plant
Docket No. 50-391

13. AVAILABILITY STATEMENT
unlimited

14. SECURITY CLASSIFICATION
(This Page)
unclassified

(This Report)
unclassified

15. NUMBER OF PAGES

16. PRICE

UNITED STATES
NUCLEAR REGULATORY COMMISSION
WASHINGTON, DC 20555-0001

OFFICIAL BUSINESS

NUREG-0847
Supplement 26

Safety Evaluation Report Related to the Operation of
Watts Bar Nuclear Plant, Unit 2

June 2013

www.ingramcontent.com/pod-product-compliance
Lightning Source LLC
Chambersburg PA
CBHW080242180526
45167CB00006B/2386